The Power of Kindness

The Power of Kindness

Why Empathy Is Essential in Everyday Life

DR. BRIAN GOLDMAN, MD

HarperCollins*Publishers*Ltd

To my partner, Tamara, and all the kind people I've met who make this world a better place.

Also by Dr. Brian Goldman

The Night Shift: Real Life in the Heart of the ER

The Secret Language of Doctors: Cracking the Code of Hospital Slang

Contents

The Power of Kindness

CHAPTER ONE

A Question of Kindness

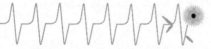

Am I a kind soul?

As an existential question, it's right up there with *Why do I exist?* and *How do I know if I'm doing the right thing?*

Up until a few years ago, the question of kindness was way down on my bucket list. I had what I believed were much more important things on my mind, such as being a competent emergency room physician. Like many of my colleagues, I worried about every mistake I made on the job, big and small. Medicine is a profession where every error carries the risk of serious consequences. I worried that my lifelong battle with insomnia, coupled with unnatural shift work, would eat into my competence. I worried about being an older physician in a young person's environment like the ER.

To those worries, add being the host of a national radio show. It's prestigious and exciting, but it's another place where getting a fact or an attitude wrong can cause instant disapprobation. Sometimes, my main comfort is knowing that even my worst mistakes on the air are not immediately hazardous to anyone's health.

At home, I was preoccupied with being a good partner to Tamara and father to my children, Kaille and Alexander. Over the years, I have focused my energies on being a solid provider to my family. For my kids, I was preoccupied with things like making sure they had the technological tools and the tutors they needed to succeed. I always remembered to encourage them to talk about their feelings, especially the ones that troubled them.

As the son of aging parents, I saw it as my duty to make sure they got astute medical care. When my mother was admitted to a long-term care facility with end-stage Alzheimer's disease, I visited her as often as I could. When she could no longer feed herself, I took turns spoon-feeding her with my sister Joanne. When my dad felt crushing bitterness at having to move his wife of nearly 60 years into the nursing home, I tried to comfort him while absorbing some of his pain.

My parents are both gone now. And that has left me with time to ask the question: *Am I a kind soul?*

From what I've told you about myself, I sound pretty kind. So how come I'm asking? Because doubt has crept into me. I have felt this way for a long time. That I'm too stressed, too busy, too preoccupied with the errors I make at work and in life, too anxious and too self-absorbed to think enough about others to be kind to them.

And I'm not alone. As I look around me, I see the same problem everywhere.

You order a decaf latte. The barista repeats your order word for word. A minute later, he's forgotten it and asks you to repeat it. When you finally pick up your beverage, one sip tells you he got it completely wrong.

A beleaguered woman struggles to lift a heavy printer onto

the service desk of a store that sells computer equipment. The printer doesn't work. The manager looks relieved when he points out that her service contract has expired.

An unaccompanied minor on a commercial flight misses a connection. For several panic-stricken hours, the child is missing. The ending is happy, but what the parents remember most about the ordeal is the call centre representative advising them not to worry.

What these and many stories have in common is lack of empathy.

It seems *everyone* has a painful story or two to tell. A recent study by University of Indiana psychologist Sara Konrath found empathy among today's college students has declined by about 40 percent compared to their peers 20 or 30 years ago, with the biggest drop after 2000.

There's sympathy and there's empathy, and many people confuse the two. Sympathy is a gesture of acknowledgement or commiseration for someone experiencing misfortune, ranging from a business setback to the death of a loved one. Think cards, flowers, or the oft-used phrase "Sorry for your loss." To extend sympathy, you don't need to know what people are going through, and you don't need to feel what they feel. Empathy, on the other hand, is the ability to use your imagination to see things from the point of view of another person, and to use that perspective to guide your behaviour.

Jean Decety, a cognitive neurosciences researcher at the University of Chicago, says that empathy is made up of several components. One is *affective* or *emotional empathy*, which refers to the capacity to feel the emotions of others. A mother who winces when her toddler trips and falls is feeling her child's pain physically. But you don't have to be a mother to experience the emotions of others. Anxious people make me feel anxious; it's as if their mood is transmitted to me like a virus. Researchers refer to this phenomenon as *emotional contagion*.

Emotional or affective empathy is an instinctive capacity that is baked into our DNA. It exists inside the most primitive centres of the brain, and it appears early in a child's life. "By the age of 12 months," Decety writes, "infants begin to comfort victims of distress, and by 14 to 18 months, children display spontaneous helping behaviors."

Emotional empathy is what motivates us to help others. On September 2, 2015, the body of Aylan Kurdi, a three-year-old Kurdish boy, washed up on a Turkish beach, and the photo of his lifeless body caused outrage around the world. Aylan was only one of more than 3 million refugees trying to flee persecution and sectarian violence in Syria. Yet somehow people who saw that one photo empathized so profoundly with the boy's fate that thousands of people donated money and sponsored refugees.

For a surgeon, emotional empathy is a double-edged sword. You want the surgeon who takes out the cancer in your chest to be motivated to help you; you do not want the surgeon to be so wracked with empathy for your pain that she hangs up her scalpel.

The second component is called *cognitive empathy*, which means having a sense of how another person is feeling. Cognitive empathy is also referred to as *perspective taking* because it involves the ability to see things from the perspective of another person. Unlike emotional empathy, which originates in the primitive brain, cognitive empathy comes from the brain's higher centres where, as Decety puts it, "motivation, memories, intentions, and attitudes influence the extent of an empathic experience." Developmental psychologists also refer to cognitive empathy as *theory of mind*, a term coined in 1978 by David Premack and Guy Woodruff to describe the ability of humans and some primates to predict the motivations and intentions of others.

Simon Baron-Cohen, professor in developmental psycho-pathology and director of the Autism Research Centre at the University of Cambridge, refers to theory of mind as mind reading. He says that humans "mind read all the time, effortlessly, automatically, and mostly unconsciously." Baron-Cohen has devoted much of his career to studying people with autism, a condition in which the ability to construct a theory of mind is greatly impaired—a condition that the eminent psychologist refers to as *mindblindness*. People with schizophrenia and attention deficit hyperactivity disorder (ADHD), and cocaine and alcohol users, may also suffer from mindblindness. They have abnormalities within the structures or the circuitry of the brain that account for this deficit.

These days, it seems that many people with decidedly normal brains behave as if they were incapable of imagining what it's like to be in another person's shoes.

Take this example: A 47-year-old woman walks up to a government service kiosk. Her husband, a recent arrival from another country, has died suddenly of a brain aneurysm. The woman wants to take her husband's body back to his birthplace, a process complicated by the fact he and his wife have applied for permanent resident status but haven't yet received it. Far from showing kindness, the clerk acts irritated. From the clerk's point of view, each encounter is supposed to take a fixed number of minutes, and this one could take an hour or more to sort out, which will drag down her productivity rating. At that moment, she seems unable to imagine the pain that the grieving widow is going through.

The third component of empathy is called *affective concern.* Some call it *compassion empathy* or *empathic concern.* Daniel Goleman, psychologist and author of the 1995 book *Emotional Intelligence,* writes that "with this kind of empathy we not only understand a person's predicament and feel with them, but are spontaneously moved to help, if needed."

Affective concern is what motivates a group of bystanders to help a person who falls in the street, or a firefighter to run toward, instead of away from, a burning building. It's what motivates a teacher to spend extra time helping a pupil with ADHD understand a challenging concept in mathematics. And it's supposed to motivate the people who care for you at hospitals and clinics: nurses, pharmacists, social workers, and doctors like me.

We expect health professionals to be competent at what they do. If they're über-competent at knee replacements or minimally invasive heart valve surgery, we might cut them some slack if they come up short in the bedside manner department.

An elderly man spends five days in a hospital corridor in the ER waiting for a bed on the ward upstairs. He is never told when the bed will be available, and his only access to personal privacy is one of two bathrooms he shares with all the other patients in the ER. A woman recovering from a double knee replacement spends the night in a urine-soaked bed—after warning her nurse to use two hands when removing the overflowing bedpan the woman depends on to relieve herself. I've heard and witnessed thousands of stories like these in health care and in the outside world. Each is different, but they have one thing in common: people in close contact with others who have no idea and little interest in what it's like to be them.

That is the very definition of a lack of empathy.

I keep referring to health professionals as if I'm talking about someone else. But I'm talking about me. There have been singular moments in my professional and personal life when I too have failed to imagine what it's like to be the other person. There's the time I didn't acknowledge the cataclysmic grief felt by a disbelieving son when I told him that his father had died, because I was worried he would accuse me of failing to save his dad. And there's the time that I grudgingly admitted an elderly woman to hospital and could not see the distress her partner felt at no longer being able to care for her at home.

I wasn't always this way. By all accounts, I arrived in this world as a smart, loving, bright, and precocious baby. I'm told I spoke in full sentences when I was 18 months old. When I attended a summer day camp as a four-year-old, my counsellor told my mother that I was like a breath of fresh air.

What became of that little boy?

In the past few years, many books have been written about empathy, both inside and outside of health care. They are helpful and important, but only up to a point. They tend to analyze empathy in terms of brain circuits and behaviours. Knowing whether or not you possess those circuits is helpful because it gives you some hints as to what gifts or burdens nature has bestowed upon you. But to expect to be more empathic by analyzing empathy is like hoping to become funnier by deconstructing jokes.

If lack of empathy is the problem, what is the cure? In his 2017 book *Against Empathy: The Case for Rational Compassion*, Yale psychology professor Paul Bloom argues that empathy based on emotion motivates people to help in ways that are counterproductive. He writes that our brains are programmed to enable us to empathize with one person at a time, which dooms us to ignore the needs of the many. Bloom says that whom we empathize with is biased in favour of those who look and act like us, which motivates us to assist people who may not be in the greatest need of our concern and our help. Instead of empathy, Bloom advocates for what he calls *rational compassion*, dispensing with emotional involvement in favour of helping others based on an objective calculation of costs, benefits, and risks.

As a call to social action, Bloom makes a compelling case. But it doesn't help me understand what I seem to be lacking as a physician. Perhaps there's a better word to describe it: *kindness*.

Michael Stein is an internist, an author, and the chairman of the department of health law, policy, and management at Boston University. In an opinion piece published in 2016 in the *Washington Post*, Stein tells the story of a woman who went to a walk-in clinic with neck pain and a fever. The woman told Stein she almost had to beg the young doctor—who was convinced the woman just had a cold—to do a test for strep throat. The test came back positive. Stein writes that the woman was angry with the doctor, calling him incompetent and lazy and accusing him of trying to save the health care system money. However, it was her assertion that the doctor was "certainly unkind" that made Stein sit up and take notice.

"It's reasonable to expect a doctor to be kind at every visit," wrote Stein. "Kindness may be less important to us when the visit is urgent, when we are in terrible pain and barely listening as we wait for relief, when the problem is diagnosed and fixed quickly. But generally, most of us assume that it matters."

On the surface, kindness sounds banal. But that may be a misrepresentation of the word's true meaning. Linguists say "kindness" comes from the Old English word *cynd*, which refers to kinship, as in friends who are "two of a kind."

"Kindness implies the recognition of being of the same nature as others, being of a kind, in kinship," wrote U.K. psychiatrist Penelope Campling in a 2015 editorial published in *BJPsych Bulletin*. "It implies that people are motivated by that recognition to cooperate, to treat others as members of the family, to be generous and thoughtful."

In other words, when I'm kind to people, I see them as being like me, and me like them.

That is how I felt when a woman named Marcela did an extraordinary act of kindness for me. On Good Friday in 2016, I was scheduled to fly from Toronto to São Paulo to do some research for this book. When I arrived at the airport, I was barred from boarding the flight because I didn't have a tourist visa. When I travelled to Brazil in 1985, I didn't need one, and the thought never occurred to me that I'd need one in 2016.

Calling this a catastrophe is an understatement. I had lined up interviews and had hired a local producer to act as my translator and to help with the arrangements on the ground. It was Good Friday, and the Toronto office of the Consulate General of Brazil was closed until the following Tuesday.

To make a long yet delightful story short, a kind official who was on call for the Embassy of Brazil answered my call to the emergency phone number that very night. He gave me the emergency number of the official who was on call that weekend for the consular office in Toronto. I called Marcela at two in the morning, and after I had explained my situation, she agreed to process my application for an on-the-spot visa by noon the next day.

Marcela knows what it means to be kind. But in much of modern culture, people (doctors included) tend to see colleagues as us and customers as them. That's hardly a recipe for being kind in the true sense of the word.

But even Campling's analysis misses what I believe to be an essential part of being kind, something that lifts it from the superficial and transforms it into a deeper experience with the potential to change the lives of both you and the person to whom you extend kindness.

I'm talking about something called *synchrony*. The word *synchrony* means "a simultaneous action or occurrence." For instance, it may refer to the matching of rhythmic behaviour between people. Synchrony is an important topic among developmental psychologists. In 1974, William Condon and Louis Sander published a groundbreaking study in the journal *Science* in which they observed the interaction between newborns and their parents. They found that as early as the first day of life, newborns move in sync with the sound of a parent speaking.

Later, Andrew Meltzoff from Oxford University and M. Keith Moore from the University of Washington demonstrated that babies as young as three days old imitate the facial expressions of their mothers. Thus, newborns mirror their parents' faces, and their parents mirror theirs. It's one of the earliest examples of what developmental psychologists refer to as *interactional synchrony*, an essential part of the process by which babies become attached to their parents.

Synchrony is also found in dance and music and in shared rituals such as chanting in church. If you have ever performed the wave at a baseball or football game, you have taken part in a mass example of behavioural synchrony. Similarly, if you've ever met someone and just clicked with them, you have experienced interpersonal synchrony. The next time you visit a coffee bar or a restaurant, watch twosomes who are seated together. It's not hard to spot those who share synchrony; they're the ones whose hand gestures and speech patterns mirror one another.

Studies have shown that people in sync have stronger social bonds. They are more likely to empathize with and be kind to one another. This is true between friends and acquaintances. What may surprise you is that it is also true between therapist and patient. A 2014 study by Zac Imel, a psychologist and researcher at the University of Utah, found that therapists get in sync with their clients. The more they adopt the speech patterns of their patients, the more they empathize with them.

Synchrony is the superhighway that leads to connection and to kindness.

You can learn something meaningful about a word by considering its opposite. The antonym of empathy is apathy. The most obvious antonym of kindness is unkindness, although meanness, churlishness, greediness, ill will, self-seeking, and malignity are also opposites.

As a doctor, I have made plenty of mistakes, most of them due to a fatigue-related failure to make the right diagnosis. As far as I can recall, I have never made a medical error out of malice. On the other hand, I'd be lying if I said I care about every patient to the same degree, at all times of the day and night, regardless of the circumstances. On a few occasions, I have made the wrong diagnosis and compounded the problem by disregarding the patient's concerns that my diagnosis was incorrect. I have told patients with injured limbs that they don't have a fracture because I failed to see one on the X-ray. In some cases, instead of taking their insistence as a cue to look again, I practically pushed them out of the ER. At these times, it was defensiveness at being accused of making the wrong diagnosis that made me tune into my own distress and ignore the concerns of the patient.

I'm not just an ER physician. I'm also a medical journalist and the host of a popular radio program on CBC Radio One called *White Coat, Black Art.* I have had a very successful career pulling back the curtain on the world of medicine by getting front-line doctors and nurses to speak candidly about their frustrations with patients, colleagues, and the hospitals where they work.

When the risk to them is clear, I try to remember to warn guests who appear on the show of the possible consequences and make sure they know what they're getting into. On at least two occasions, I failed utterly to do so. In the first instance, I let the physician tell his story even though he broke at least one hospital regulation in doing so. In the second instance, I put the comments of another physician on the air even though the doctor had not given his permission to do so. I talked myself into believing that I could protect the doctor's identity by not mentioning his name on the air. It turns out his voice was instantly recognizable by many people, including those with power to cause him all kinds of grief. While their jobs were never in jeopardy, both got into serious trouble with their bosses. In both instances, my interest in enhancing my broadcasting career won out over their welfare.

In the ER, I have been competent yet on occasion unkind. In my broadcasting career, I have sometimes put personal ambition ahead of the interests of the people I interviewed.

Over the years, there have been many other examples that

I have been able to shake off. But some are different. That's because I subsequently found out how much my actions hurt the people involved. Each forced me to see things empathically—that is, from the other person's point of view—albeit in brief glimpses that I found painful to reflect upon.

Nobody likes to be criticized. I tried not to dwell on these moments, because each made me feel ashamed. The incidents I found out about forced me to see myself as those affected saw me: selfish, unempathic, and unkind.

For years, I have carried and accumulated these stories and the regret I feel about them as if they were the price of doing what I do, with nothing to be taught and nothing to learn.

I need to know why I can't be kind more often. Was I born without the wiring inside my brain? Or did I have it and lose it? Am I too busy, too selfish, too stressed, too preoccupied with self-doubt about my clinical skills and other abilities? If it's in my nature and in the hard-wiring of my brain to be kind and empathic, then how do I get these qualities back?

My curiosity about kindness has taken me on a voyage lasting close to two years. I have travelled across Canada and the United States and as far away as the United Kingdom, the Netherlands, Brazil, Japan, and Australia. My aim was to meet the kindest and most empathic people on the planet, to hear and share their stories, and to learn what makes them extraordinarily kind. A few work in health care, but most toil in places like bars, fast food restaurants, and call centres. Some are experts in neuroscience who peer inside the human brain. Others are at the leading edge of efforts to simulate human empathy and kindness in robots and androids and in the virtual world of computer games. All have helped me understand what it means to be kind.

This is a journey into empathy as witnessed through my eyes. I'm your tour guide. Sometimes I'm the guinea pig for a test of empathy, sometimes a vessel for you to experience and learn how to be kinder in a world that could use a bit more kindness.

Although I practise medicine in an ER and talk a bit about my experiences herein, this book does not provide medical advice. I have changed some names and descriptions.

This is not a book about me. It's a book about us.

CHAPTER TWO

Hard-wired

I enter the Centre médical Mailloux just after lunch. Except for the signs in French, the two-storey, flat-roofed structure made of red brick looks like any other medical clinic in a strip mall. Located in suburban Quebec City, it has a fairly typical complement of family doctors, physiotherapists, an audiologist, a walk-in clinic, and an acupuncturist. This seems like an unremarkable place to find out if my brain is hard-wired to be empathic and kind.

The doors open on the second floor, and I make my way to room 225 and a door marked IRM Québec, which stands for Imagerie par résonance magnétique (magnetic resonance imaging, or MRI). IRM Québec operates four private MRI clinics in the province. I still think of MRIs as rare and exotic diagnostic images. Here, at this utterly pedestrian clinic, it's scans for hire. Seven hundred and fifteen bucks buys you an MRI of the abdomen or prostate. "Articulations ou sacro-iliaques" (joints or the sacro-iliac) will set you back $665, the same price as a scan of the head.

"Monsieur Brian Goldman," a middle-aged woman says and points to a cubicle.

"Retirez toutes les clés et pièces de monnaie de vos poches. Votre bague de mariage aussi." The woman instructs me to remove keys, coins, and other metal from my pockets and to take off my wedding ring. That's because the magnet in the MRI scanner heats up metal and can burn your skin.

I exit the change cubicle and walk to the vestibule just outside the MRI scanner. A thin young man is holding a clear plastic box filled with plastic goggles with prescription lenses.

"You need to take your glasses off," says Mathieu Gregoire, a recent PhD in psychology at Laval University in Quebec City. Laval has become a Mecca for brain research on empathy. Gregoire is the lead investigator in a study measuring empathy in health professionals and comparing the results to a group of lay people. I'm one of his preliminary test subjects.

A man in his forties grabs a pair of goggles from the box Gregoire is holding. He takes his own glasses off and looks through them. "Here, try these on," Philip Jackson says as he hands me the goggles. I find them good enough to read text on a nearby TV screen.

"You could have been an optician," I reply, as I adjust the elastic strap.

"Good," he says. "You need to be able to see clearly the faces that we show you."

Jackson is a professor of psychology and neuroscience at Laval and one of Canada's top researchers in empathy. He was Mathieu Gregoire's supervisor on his recently completed doctoral thesis and is collaborating with his young protege on the study in which I am participating.

Before coming to Laval, Jackson did postdoctoral studies at the University of Chicago. There, he co-authored seminal papers with Jean Decety, a heavy hitter in empathy research. Jackson and Gregoire are working on a hypothesis that might go a long way toward explaining the apparent lack of kindness among health care professionals.

"Part of my work that I've done with Philip Jackson is about the impact of being repeatedly exposed to the pain of others," says Gregoire. "Research done by others has shown that for health care professionals, being exposed on a daily basis to the pain of others leads them to underestimate the pain of others. Compared to a control group of people not exposed to others in pain, their brains are not reacting in the same way."

In my 35-plus years in the ER, I've seen thousands of patients in horrible pain, everything from broken hips to kidney stones to heart attacks. Taking care of their pain is part of my job. I treat their pain, but I don't tune into the abject fear and despair that they feel.

On the other hand, for non–health professionals—loved ones and bystanders alike—seeing someone screaming in agony has an emotional impact. Experts like Philip Jackson and his mentor Jean Decety call that experience *empathic resonance*, a term that refers to a heightened state in which one person tunes into another person's feelings. Jackson says that when a non-medical person experiences empathic resonance, specific parts of the brain light up. Researchers can identify these on a functional magnetic resonance imaging brain scan, or fMRI. The scan measures the activity in various parts of the brain by detecting changes in blood flow.

I'm about to have my very first fMRI scan of my brain. There's no limit to what I'm prepared to do to see where my kindness has gone.

"We'd expect to see activation in areas involved in the pain [processing parts of the brain]: for example, the cingulate cortex and the insula," says Jackson. He predicts that my fMRI brain scan will show reduced activation in both the cingulate cortex and the insula, a finding he and Gregoire have already demonstrated among health professionals.

"In a study we ran on nurses in Halifax with our colleague Margot Latimer [from the school of nursing at Dalhousie University], part of this reduced activation was associated with the number of years of work experience of the nurses. The greater the number of years they worked, the lower the activation on the brain scan."

That paper—which has been submitted for publication in a peer-reviewed journal—is one of the first experiments to show that the capacity of veteran nurses to respond to the pain their patients feel is dulled by years of experience.

Jackson stressed that he and Gregoire know *what* appears to be going on inside the brains of veteran health care workers. They just don't know *why*. "It could be due to frequently having to treat lots of patients in pain," says Jackson. "But it could be something in the health professional's environment, the work, the stress, and other factors. One initial interpretation is that health care professionals are either habituated to patients in pain so they respond less. Or it could be that they deliberately distance themselves from the pain of others. We're not sure which it is, or if the two can be teased apart, but that's what we would like to show in future studies."

Gregoire is quick to point out that there may be good reasons why veteran physicians like me underestimate our patients' pain. "Every time we talk to physicians, they're telling us that maybe they underestimate pain in order to not give too much opioid pain relievers," Gregoire says. "To make sure they don't overdose the patient on opioids."

That makes sense. As a physician, I can't be kind just for the sake of being so. Every decision I make is weighed against the risk of doing harm. Opioid overdoses are a leading cause of death for young people in Canada and the United States, and doctors' prescribing habits are the major reason.

When I assess a patient with a broken bone, I mentally compare him or her to many other patients with similar injuries. Some are stoic and others complain bitterly. I don't want to underestimate a patient's pain. But the stakes for overestimation can't be ignored.

Gregoire hands me a questionnaire to fill out. It's to make sure I don't have any metal implants in my body and that I'm not prone to claustrophobia; both are MRI deal breakers. As I sit down to fill out the paperwork, I can hear the faint but pulsating thrum of the MRI machine on the other side of the glass.

"The scanner is safe for the body," says Gregoire. "The danger is getting metal inside the scanner, so that's why we want to make sure you don't have any small bits of metal on you. A piece of metal can actually move in the scanner or heat up inside your body. If you bring in a pen, it can fly inside the scanner."

Documents complete, Gregoire brings me up to speed on the experiment in which I'm about to be a subject. "We'll give you earplugs because it's really noisy inside," he warns before explaining what I'll have to do while my brain is being scanned. "The task that we're going to do today is called a pain estimation task," he says. "We're going to show you different faces of people in pain, and you'll have to rate the level of pain on the person's face on a visual analog scale."

A visual analog scale, or VAS, is a standard instrument used by pain researchers to measure pain. Basically, it's a horizontal line marked by no pain on the left and maximum or worst pain imaginable on the right. My job in the scanner will be to use a joystick to move the cursor to the place on the line that best represents how much pain the person in the photograph is feeling. I'll have to be fast on the trigger. "After four seconds, the image is gone," says Gregoire.

We do a practice round in the vestibule to make sure I can do it at top speed. Then, the young researcher runs through the experimental design.

"First, it's going to be seven minutes of doing nothing while we take a baseline scan of your brain," says Gregoire. "And then, we're going to do three runs of nine minutes of photos where you rate the pain."

The young scientist will be looking carefully at my fMRI brain scan to see what happens to my cingulate cortex and insula, the two areas that light up when a layperson sees someone in severe pain.

A technician leads me into the scanning room. Before me is a thick cylinder about three metres in length. In the middle of the cylinder is an opening with a platform a little wider than my torso. The structure has a futuristic appearance.

I climb onto the MRI platform, and the technician guides my head into a helmet-shaped contraption called a head coil. The coil acts as an antenna to receive radio frequency signals coming from my body and transmits the signals to a computer that turns the data into a picture of my brain. She attaches a mirror to the coil so I can see a reflection of the photos of people in pain that are being projected behind me.

"Everything okay, Dr. Goldman?" Mathieu Gregoire's voice on the intercom jars me.

"Everything's good," I reply. I'm lying. I don't suffer from claustrophobia, but I've had an anxiety attack or two in my life.

The coil feels tight, and I don't like the idea that I must not move. I breathe slowly and deeply to try to control the urge to run from the room.

"Please remember to stay still," says Gregoire. "Try not to move your feet. If you have to move your hand, do it now before we start the task."

Even with earplugs in place, the pulsating sound from the fMRI's huge magnet is deafening. The first projected image appears on the mirror attached to my helmet coil. It's a middle-aged man who is wincing in pain. Is he in a lot of pain or a little? I've got four seconds to pick a spot on the pain line. I'm finding it harder than I thought to work the joystick with my left index and middle fingers. I undershoot, then I overshoot.

With time running out, I arbitrarily choose. I'm still thinking about the middle-aged man as the second image pops up. An old woman. Is she grimacing or not? This is tougher than I thought it was going to be.

I should ace this little experiment. But that's false bravado. The truth is, I have no idea how I'll do any more than I know how caring I am these days. It's strange to think that the fMRI images of my brain will tell Gregoire and Jackson things about me that I don't know about myself.

Still, if they and other leading neuroscientists are correct, the search for human kindness begins inside the brain.

It's no accident that my first stop on the road to kindness is the brain. Neuroscientists believe that humans are hard-wired to be empathic and kind. They have identified several parts of the brain that appear to be important for empathy. The cingulate cortex and the insula—structures mentioned by Philip Jackson and Mathieu Gregoire—are two of them. The anterior insula seems to be very important. In 2012, researchers at Mount Sinai School of Medicine in New York and colleagues in Beijing found that patients whose anterior insula had been damaged by brain surgery were unable to read faces like the ones Gregoire showed me. In other words, they had lost the ability to empathize with people in pain.

In 2013, researchers from the Max Planck Institute for Human Cognitive and Brain Sciences reported that we lack empathy when a part of the brain called the supramarginal gyrus doesn't function properly. This part of the brain is important in helping us compare our own emotional state to that of other people.

Christian Keysers, co-director of the Social Brain Lab at the Netherlands Institute for Neuroscience in Amsterdam, is one of the leading lights of this kind of research. Like Philip Jackson, Keysers is trying to map kindness inside the human brain.

Keysers and Valeria Gazzola (Keysers' collaborator who also happens to be his spouse) study the circuits of the brain that get activated when we do an action, experience a sensation, or feel an emotion, and when we watch someone else doing or experiencing the same thing. The couple belongs to a group of powerhouse brain researchers that includes Stanford University's Jamil Zaki, who studies the brain circuits involved in motivating people to engage with or avoid other people's emotions, and Daryl Cameron, a researcher at the Iowa Morality Lab who studies what induces people to be kind to one another.

Keysers' 2012 book *The Empathic Brain*—which won the Independent Publisher Book Award for best science book— cemented his reputation as a world-class researcher with a knack for making brain science compelling and easy to understand. In the book, Keysers tells the story of the accidental discovery of a set of brain cells that are critical to the hard-wiring of empathy and talks about his role in advancing this area of research.

When you perform simple actions like curling your fingers to grasp the handle of a cup of coffee or playing a scale on the piano, a precise group of cells in your brain switch on so they can control the muscles in your fingers to orchestrate the action. This much researchers already knew. The discovery that Keysers writes about is that the very same brain cells that switch on when you move also switch on in the same way when you *watch* someone else move in the same way. Keysers calls the unexpected finding "as surprising as discovering that your television, which you thought just *displayed* images, had doubled all those years as a video camera that *recorded* everything you did."

In August 1990, a team of neuroscientists at the University of Parma in Italy was testing the brain activity of a macaque monkey. The scientists—Leonardo Fogassi, Vittorio Gallese, and Giacomo Rizzolatti—had inserted a tiny electrode the diameter of a hair into the monkey's brain. One end of the electrode had been placed on a single brain cell in a part of the brain that controls movement; the other was connected to an oscilloscope to display the electrical activity of the cell.

The researchers offered the monkey a raisin. The monkey grasped the treat, and the oscilloscope dutifully recorded a green burst of spiky discharge, a visual representation of the monkey's brain recognizing the food and then activating the monkey's paw to grab it. It was a completely ordinary confirmation of what scientists would expect to be going on inside the primate's brain.

But what happened next was extraordinary.

Vittorio Gallese just happened to grasp a raisin in front of the monkey, and the oscilloscope recorded exactly the same kind of spiky discharge as happened when the monkey itself grasped a raisin. At first, the researchers assumed that the discharge was interference or a glitch in the machine. Once they ruled that out, they realized that the cell the electrode was record- ing inside the monkey's brain had a special property that had never been observed before: a brain cell that was activated *per- forming* an action and *watching* someone else perform the same action—as if someone were holding up a mirror so the monkey could see itself performing the action.

The researchers dubbed the tiny brain cell a *mirror neuron*.

Keysers was finishing his master's degree at the University of Konstanz in Baden-Württemberg, Germany, when the Parma scientists made their discovery. When he was finishing his doc- torate at the University of St. Andrews in Scotland, he attended a lecture by Vittorio Gallese and had an epiphany on hearing about the new discovery.

He was so impressed that he joined Gallese and colleagues in Parma. He knew from the start that mirror neurons would guide his career. "Later on, I was in the lab and then I witnessed my first mirror neuron myself," says Keysers in an interview via Skype. "First of all, it's a surprise because you wouldn't expect the motor cortex of the brain to respond to *observing* what other people do. Then of course your next instinct is to say, 'Wow.'"

The implications of the discovery of mirror neurons were staggering. Until that point, scientists believed that brain cells could serve only one function: In the example of picking up a raisin with the fingers, they believed that there are brain cells that see the raisin and different brain cells that move the fin- gers to grasp the raisin. Orchestrating the grasping of the rai- sin required that two different parts of the brain communicate with one another.

"The discovery of mirror neurons changed this view of the brain's division of labor," wrote Keysers in *The Empathic Brain*. "Mirror neurons have a dual purpose, both perceiving the world and acting on it."

Mirror neurons don't see *or* do actions, they see *and* do actions.

To Keysers, that dual purpose offered a tantalizing way to explain the brain's role in making humans kind and empathic. As I've explained, cognitive empathy is the capacity to imagine things from the perspective of another person. Having mirror neurons that fire up when we see someone else perform an action suggests that humans are hard-wired to make a connection between the other person's actions and our own, an ability that would come in mighty handy for a species that is programmed biologically to empathize with others.

If lesser creatures on the evolutionary scale have something as sophisticated as mirror neurons, it makes sense that humans would have them too. Still, believing humans possess the same equipment and proving its existence inside living human brains are two different things.

The challenges facing Keysers and others are twofold: prove that humans possess mirror neurons, and figure out exactly how these brain cells make humans empathetic.

Proving the existence of mirror neurons in humans is not as easy as it sounds. You can't insert electrodes inside the brains of living people to satisfy scientific curiosity. So, Keysers and his colleagues at the Social Brain Lab have led the way in developing techniques that indirectly measure brain activity in humans without having to insert electrodes. They use electroencephalograph (EEG) readings to map out in real time how different parts of the brain talk to one another when a human subject performs a task or observes the task being performed by someone else.

And at UCLA, neuroscientist Marco Iacoboni uses fMRI to generate images of the brains of test subjects. Again, as in the monkey experiment, Iacoboni has found that these images look the same when test subjects perform a movement and when they watch someone else do the same movement.

To Keysers, the only difference between fMRIs on humans and the monkey experiment is the degree of pinpoint accuracy. "When we record signals from neurons in monkeys, we know that a single neuron is involved in both doing the task and seeing someone else do the task," Keysers told the American Psychological Association in an article published in 2005. "With imaging, you know that within a little box about three millimeters by three millimeters by three millimeters you have activation from both doing and seeing." That little box contains millions of neurons. Keysers uses an ultra-refined version of fMRI to reduce the size of the box from cubic millimetres to cubic micrometres.

"We now have rock solid evidence for the fact that the motor system gets reactivated when humans see the actions of others," says Keysers. "So, the phenomenon for actions is fully described."

What Keysers and other researchers have established is that humans and monkeys possess mirror neurons that switch on when performing an action or observing someone else doing it. Showing how mirror neurons make us empathic is a lot more complicated.

"Figuring out the connections between microscopic brain cells is like opening a watch and trying to figure out how the mechanism tells the time," says Keysers.

And once they map out the connections, the task is to figure out how the mechanism guides us in our everyday lives—

everything from deciding whether to hold a door open for a stranger to how to raise our kids.

Keysers is studying how mirror neurons help explain the link between empathy and emotion. When we are kind to others, we are hard-wired to feel happy. Keysers went looking for a different set of mirror neurons in the brain, ones that by design are connected to our emotions automatically: for example, disgust.

In 2003, he and Bruno Wicker of Aix-Marseille University published a study in which they took fMRI images of the brains of 14 male recruits as they sniffed butyric acid, which smells like rancid butter. The assault on their sense of smell caused the anterior insula to light up.

In the next part of the experiment, the study participants weren't given butyric acid to sniff themselves. Instead, they watched a film in which a trained actor looked like he was smelling something foul, complete with a facial expression showing disgust. On the fMRI scans, the anterior insula again lit up. By proving that experiencing disgust or observing it in someone else activates the same part of the brain, Keysers and Wicker had demonstrated the existence of mirror neurons that are attached to an emotional state.

Keysers proved it yet again with an experiment involving touch. He recruited volunteers who were lightly touched with a feather, after which they watched images of someone else being touched in the same manner. As in the experiment involving the rancid smell, Keysers proved that being touched and watching someone being touched activated the same part of the brain.

Keysers calls this finding *tactile empathy*.

You can see where he's going with this research. If your brain reacts to someone else being tickled as if you are being tickled yourself, on some level, you must be empathizing with the person being tickled. And if that's the case, then our brains are hard-wired to be empathic.

That's the hypothesis. More than two decades after the discovery of the first mirror neurons in monkeys, Keysers admits that research connecting mirror neurons to empathy lags far behind.

"We see that some of the brain regions that are active when we experience emotions also get activated when we witness the emotions of others. What we don't know is whether it's the same brain cells that get activated," he says.

Assuming he can figure that out, the next question is even more challenging.

"We know we experience empathy, and we know we activate these brain regions. But does that mean that we know that it is the activation that causes the empathy?" he asks.

The mirror neuron theory has attracted its share of critics. They point out that studies have shown that these brain cells are also found in some of the most predatory species in the animal kingdom. Critics also argue that the concept is too simplistic to explain why people are empathic.

Keysers agrees that empathy can't be explained by mirror neurons alone. He thinks there has to be something more: "My feeling is that we probably have two mechanisms at play that interact with each other."

He has a hypothesis that might help explain why empathy is in short supply inside and outside of health care, and that might just help me on my personal search.

As I wrote at the beginning of this book, empathy or kindness is part instinct and part choice. If a person collapses in the street, most of us have an impulse to offer assistance. Some have a strong urge and others less so. The people I'm meeting for this book are on the caring side of the spectrum. That said, the tendency to help probably follows a bell curve. Helping a person in distress is known to psychologists as *prosocial* behaviour because doing so is good for society. Without it, we would have a hard time getting along with one another.

Keysers believes that mirror neurons provide humans with an instinctive and instantaneous capacity to empathize with someone in severe pain or disgusted by a noxious smell. At his lab in the Netherlands, Keysers and his team conduct studies in which they perform brain scans on volunteers as they watch films. "Whenever we put someone in the scanner watching a movie, we observe a certain degree of empathy without having to build the specific context that would make the person relate to the person in the movies," says Keysers. "It is not strong, but it is present relatively quickly and is maintained as a kind of default state."

That would be your basic, built-in vanilla capacity to be kind. It's there, it kicks in automatically, and it's not very strong. But humans don't operate solely on instinct. They balance altruism with self-interest.

"You see a homeless guy on the street," says Keysers. "You only have $1 left in your pocket that you had reserved for an important purpose. What will you do? The evidence shows that you will cross to the other side of the street, and look away from the homeless person, all to make sure that your empathy does not get in the way of what you had meant to do with that dollar."

How that happens is where Keysers' theory gets compelling. "Over the past few years, I have come around to the idea that there is a very elaborate executive system in your brain that tries to make sure that the kind of empathic ability that I've been researching is not used in a maladaptive way," says Keysers.

Keysers suspects that humans have an empathy "off-switch" that for most of us is strong enough to override the instinct to be kind. In between the two systems, the brain's higher functions are busy calculating risk and reward. "Whenever you foresee that empathy will cost you resources, you have something in your mind and brain that directs your attention away from empathy so that you save the cost," he says.

You empathize only when you calculate that there is a personal gain in it for you. And the gain, says Keysers, is not always material: "The gain might be the pleasure that you take out of being empathic."

Keysers believes that the model of empathy he's formulating would go a long way toward explaining the behaviour of health professionals like me. "If you're a surgeon, you have to work day after day," he tells me, "and you're going to see terminally ill patients. The cost may be burnout, or it may be an inability to operate if you are burdened by the suffering of your patients. You can regulate those costs by not attending to the emotional needs of the patient whose suffering might burden you."

If you're disturbed by the thought of an uncaring surgeon, it may be a matter of emotional and professional survival. How could a surgeon operate if he or she worried about how much pain the patient feels following the operation?

Keysers has a provocative theory about that, too. "Those who gravitate toward surgery are those who can manage to regulate empathy well enough not to be repelled entirely by that line of work," he says.

In other words, surgeons have a really good empathy off-switch.

* * *

My fMRI brain scan is completed, and my ears are still ringing from the experience. When doctors talk about the scanner's side effects, they always mention claustrophobia. I've never heard them mention ringing ears.

Philip Jackson drives me to his lab to show me what else he's working on. "They're copying the data from your scan onto a USB key so we can analyze it," says Jackson. "Did you have any problems following the instructions?"

"I got a little tired during a part of it," I admit. "I may have missed one or two of the faces."

Jackson says they speed up the test for budgetary reasons. "We give you just four seconds to answer because we pay by the minute here," he explains. "So people have to think and respond fast."

Along the way, he confides that he has some serious misgivings about fMRIs as instruments to measure empathy. "I've trained in it, I still use it, and I believe in it," he says. "But I think that a lot of people represent it in the media as a miracle machine, some sort of mindreader. It's just a signal or correlate of brain activity or what's going on in the brain and it's very crude."

Crude *and* simplistic. Jackson says fMRI scans show neuroscientists which areas of the brain light up when research subjects are being tested for empathy. But, as Keysers said, the scanner can't tell researchers how different parts of the brain communicate with one another. The most misleading aspect of fMRI scans is that they happen in a sterile imaging room far from everyday life. "You're still lying down inside a very loud machine," he says. "The machine is not a normal environment."

What Jackson means is that human empathy takes place in the real world. The shopkeeper who notices that a child looks sad and hands out a cookie to cheer her up. The call centre representative who hears the distress in a customer's voice and takes him patiently step by step through a reboot of his computer. In both instances, a gesture of empathy or kindness is extended: the product of a unique interaction between two people. The details of the interaction are so different from any other that it's difficult to make generalizations.

As we pull into the parking lot of the Centre de recherche de l'Institut universitaire en santé mentale de Québec (CRIUSMQ), a 10-minute drive from Centre médical Mailloux where I had my brain scan, Jackson is about to show me a brand new way to measure empathy. He's building a computer avatar with a human face that displays emotions and measures precisely how test subjects react, including heart rate, respiration, oxygen level, skin temperature, and eye movements.

Jackson takes me to his lab, where three of his graduate students crowd around two large, rectangular computer monitors. On the right-hand screen is a close-up of a computer-generated yet distinctly human face, with dark hair and dark eyes.

"It's a woman," I say out loud.

"She needs a bit more polishing," says Jackson.

A graduate student named Daniel is at the controls. He switches the avatar to a man around 35 years of age with light brown eyes, thick eyebrows, and full lips that look a bit pouty. Wait. Did I just see his lips move?

"He's getting kind of angry," I blurt.

"Yes," says Jackson. "Or, he could be in pain."

I don't know which it is, but the avatar sure is lifelike. It's only then that I notice Daniel working slider controls with gauges on the left side of the computer screen. Different sliders control different facial muscles.

"Do you see the brow?" Jackson asks. "One brow is going up. Perhaps we can show the mouth moving." The avatar's mouth moves into a smile so realistic I can't help but smile myself.

Jackson catches my expression and grins in triumph. "This is just to show how individual and how independent these features are," he says. "We can program the avatar to display whatever emotions we want—expressions like happiness, sadness, pain, and so forth."

"Disgust?" I ask.

"We actually know all the micro-movements that add up into an expression of disgust," he says.

I have so many questions. I start with, "Is he supposed to look like somebody?"

Jackson laughs. "The first version of the avatar looked too much like me, so I asked to change the face," he says. "I specifically wanted to avoid—how do you say it?—a narcissistic role."

The avatar is called the Empathy-Enhancing Virtual Evolving Environment—or EEVEE for short. If the name sounds familiar, it should. Eevee also happens to be the name of a character who first appeared in the video games Pokémon Red and Pokémon Blue, and is so named because its unstable genetic code enables it to evolve into a number of different Pokémon characters (in Japanese the name is *Eievui*, which means "evolve.")

Jackson's avatar is controlled by an intriguing software program: the Unity game engine (developed by Unity Technologies). He and his team programmed "eye tracking, face reading, physiological response based on heart rate or something else so that EEVEE can listen to these signals." He continues to describe the avatar's qualities: "So the idea behind EEVEE is to build an avatar that's able to detect physiological response from a human and able to get interaction with that human and take into account those physiological responses to modulate the interactions."

Jackson shows me how EEVEE can track a research subject's eye movements. The focus of the subject's eyes is projected as a red tracking dot on the computer screen. Jackson gets Daniel to demonstrate. "The tracking dot shows where Daniel is looking," says Jackson.

In other words, the avatar can tell if a human test subject is making or maintaining eye contact.

Jackson says he's developing the system as a training tool for, say, health professionals. The team is working on an exercise in which the avatar grimaces in pain, and the program adjusts the grimace depending on how much the student shows empathy for the avatar.

The avatar Jackson shows me is a white male in his thirties. Soon, the range of avatars will grow. "We'd like to manipulate the age of the avatar by changing the skin and so forth," he says. "It could be the age, the colour of the skin, and the texture."

To gain wider acceptance of EEVEE, Jackson needs to gather lots of data. But the results are promising. In a 2015 study published in the journal *Frontiers in Human Neuroscience*, they demonstrated that EEVEE avatars were as convincing as trained actors.

Jackson says he wants to use EEVEE to study empathy in normal human subjects, as well as in those who suffer from conditions in which empathy is reduced, such as autism spectrum disorders (ASDs) and schizophrenia. Jackson believes EEVEE can be used to train subjects to increase eye contact with others, and to use facial expressions and a tone of voice that demonstrate concern.

"A few years down the line, EEVEE might be able to identify and make explicit behaviours that show someone being empathic," he says.

There's a creep factor to computer-driven avatars teaching humans how to care, and Jackson knows it. "We want to build sensitive or empathic avatars, not to put avatars ahead of humans," he says. "To understand what makes human beings more or less empathic, what are the markers of this empathy, and what is the path perhaps toward changing or improving empathy in humans."

Avatar teachers of empathy aren't replacing humans just yet; nor is EEVEE replacing the fMRI. Jackson sees them being used in tandem. "Functional MRI is one of the many tools we have in cognitive neuroscience to have a perspective of brain function," he says.

"Do you still have ringing in your ears?" Jackson asks as he drives me to Jean Lesage International Airport in Quebec City.

"I think it's getting better," I lie, not wanting Jackson to feel bad.

"It can be pretty strong," he says.

I keep pulling on my left ear trying to make the noise disappear.

"We'll ask Mathieu Gregoire to run the same type of analysis he ran for other subjects," says Jackson. "In the study in which you observed facial expressions of pain, we'd expect to see activation in different areas of the brain involved: for example, the cingulate cortex and anterior insula. That might take a few weeks."

Just hearing that it's going to take weeks and even months for Gregoire to analyze the data from my scan makes me feel anxious. What if he finds that caring parts of my brain are damaged or missing?

Jackson catches my look of concern. He tries to reassure me.

"You can only measure empathy by what you do," he says. "I think people can do the right thing without feeling it and without caring. So, is it necessary to be empathic? Do all the people working in a hospital need to be empathic the same way? If the job is done, if everyone is happy and the patient is better, then I believe not everyone who works there needs to be empathic at the same level. That's my feeling."

I mull over Jackson's words as I sit in the airport departure lounge and wonder if he's right.

In the meantime, I won't be sitting around waiting for the fMRI results. I've got some extraordinarily kind people to meet. Perhaps they can show me what I'm missing. But first, I'm off to Western University for psychological testing.

CHAPTER THREE

Psychopaths, Narcissists, and Machiavellians—The Dark Triad

"I've attached a couple of questionnaires," writes Derek Mitchell in an email. "Your results will do nicely for your objectives. I will have someone in my lab score them tomorrow. I look forward to meeting you."

My objectives are the same ones that took me to Quebec City to have Philip Jackson do an fMRI scan of my brain: the same ones that are sending me across North America and around the world. Call it the quest to recover my kindness quotient—the thing I seem to have misplaced somewhere along the road to becoming a veteran ER physician.

Mitchell is a neuroscientist at Western University in London, Ontario. He's interested in how the brain is wired up incorrectly for empathy. His passion is studying people who for one reason or another aren't kind at all.

There are three broad types of people who make the list. The first are narcissists. Narcissistic personality disorder is a condition in which people have an inflated sense of their own importance, a deep need for admiration, and a lack of empathy for others. The second are psychopaths and sociopaths. Psychopaths are defined as being amoral, extremely egocentric, unable to love or establish meaningful personal relationships, and unable to learn from experience. Sociopaths are defined by lack of empathy toward others coupled with amoral conduct and an inability to conform with the norms of society. The third category is Machiavellianism, a disorder of people who are so focused on their own interests that they manipulate, deceive, and exploit others to get what they want.

According to a 2014 paper published in the journal *Personality and Individual Differences*, the common denominators to each of these states of mind are "feelings of superiority and privilege . . . coupled with a lack of remorse and empathy . . . [that] often leads individuals high in these socially malevolent traits to exploit others for their own personal gain."

In 2002, a couple of psychologists at the University of British Columbia named Delroy Paulhus and Kevin Williams lumped narcissists, psychopaths, sociopaths, and Machiavellians into what they called the "dark triad." Dark triad people are severely challenged in the empathy department. Mitchell's email contains two attachments, one of which is a test to determine whether I have any dark triad tendencies.

I click on an attachment called the PPI-R, which stands for Psychopathic Personality Inventory—Revised. It's a five-page questionnaire with 154 statements. The PPI-R is used on criminals and non-criminals alike. The test can spot if you're trying to disguise your personality traits. For instance, it asks about the same personality trait more than once, in several different ways, to see if your answers are consistent. The PPI-R can tell if you have psychopathic personality traits. It *can't* tell if you're a criminal.

I look over the test. It asks you to decide how false or true each statement is in describing yourself. The options are *F* for *False*, *MF* for *Mostly False*, *MT* for *Mostly True*, and *T* for *True*. The instructions tell me to answer as honestly as I can.

"If I really want to, I can persuade most people of almost anything," I read aloud the first statement.

Some statements are easy to answer, like number 50: "I am high-strung." Reading that one makes me chuckle. I mark that one *True* without a moment's hesitation. I'm sure my family and colleagues would agree.

Some, like one about worrying if I have hurt the feelings of others, give me pause. After thinking for five minutes, I mark that one as *MF* for *Mostly False*. I keep thinking of my answer as I work my way through the remainder of the test. I hope I'm not fooling myself.

Test complete, I click on the second attachment. It's called the AQ, short for the Autism-Spectrum Quotient. The test was developed by prominent autism researcher Simon Baron-Cohen. The 50-statement AQ questionnaire is designed to tell whether adults have symptoms of autism spectrum conditions.

"Below are a list of statements," I read aloud the instructions at the top of the page. "Please read each statement <u>very carefully</u> and rate how strongly you agree or disagree with it by circling your answer."

My eyes scan down the page to the first statement: "I prefer to do things with others rather than on my own." I sigh out loud. I *prefer* to write by myself. I *prefer* to go to films with my partner Tamara. I *prefer* jogging alone. I *prefer* working with colleagues in the ER to working alone in an office. This one stumps me a bit.

The statements get easier.

"When reading a story, I find it difficult to work out the

characters' intentions." I smile to myself. I'm always figuring out what's in the minds of characters in books, plays, and films.

Half an hour later, I'm done. I scan the answer sheets for both tests and email them to Derek Mitchell. Then I go to bed.

The next day, I try to put the tests out of my mind as I drive along Highway 401 west to London. I'm second-guessing my responses to the statements in the PPI-R. Did I answer honestly? Or did I try to game the test by picking what I thought was the morally correct answer?

I started this journey by admitting that there are days when I don't see myself as a kind person. What I fear most is that Mitchell will see something in my test scores that I can't, and perhaps don't want to, see about myself.

The Brain and Mind Institute at Western University is an internationally respected centre for the study of how the brain enables us to think, remember, plan, theorize, and feel.

I find a parkade and walk a short distance to the Natural Sciences Centre, the site of Derek Mitchell's office and lab. Like Christian Keysers, the mirror neuron scientist from the Netherlands, Mitchell studies the circuits of the brain involved in empathy. But he's chosen an intriguingly peculiar path to discovery. I'm on a search for the most empathic humans on the planet. Mitchell, on the other hand, has gone in the opposite direction. He's mapping out the brain circuits for kindness by studying people who by dint of injury, disease, or mental health condition lack compassion.

Mitchell greets me with a warm handshake and a smile. As we start to chat he seems warm—and empathic. I ask if he perceives himself that way too.

"If you ask my mom, she'd say I've always been a sensitive boy," Mitchell says. "I think it was just something that came early and naturally for me. There are times where people tell me that I can be too sensitive, so it's not always an advantage."

Whatever it is that makes Derek Mitchell a kind researcher, it doesn't run in every member of the family.

"My brother, for instance, who was raised in the same environment, is not what my mom would call a sensitive boy," says Mitchell. "I do think that there are a number of genetic components, but I'm not suggesting there's something predetermined here."

The nature-versus-nurture aspect of empathy has been a big part of Mitchell's career from his early student days at the University of British Columbia. He almost didn't become a neuroscientist at all.

"When I started, I was really interested in English literature and character studies," says Mitchell. "I loved Dostoyevsky, and I realized what drew me to literature were the characters, what you can learn about people and what literature teaches us about human nature and emotions." Interesting. For readers to connect with fictional characters, a novelist must make it possible to empathize with them.

But literature wasn't to be Mitchell's field of study, thanks to a chance series of events that propelled him toward a career in psychology. "I remember trolling around the psychology department one day," he recalls. "One of the PhD students invited me to volunteer to work on a study of fear and anxiety."

Later Mitchell got invited to work on a psychology study of incarcerated people. Others would have recoiled at the offer; Mitchell was thrilled. "Before I knew it, I found myself in downtown Vancouver at 5 A.M. waiting for the grad student to pick me up so we could head off to the jail to interview people," he recalls. "I came to the conclusion that this is a way of studying empirically the same things that fascinate me about English literature."

Mitchell threw himself into the psyches of incarcerated people. One case in particular piqued his interest and launched his career.

"I interviewed someone my age," Mitchell recalls. "He came from a very good family that was very affluent. For whatever reason, he started developing these horrible behavioural problems around puberty. He was very violent. He would steal his parents' cars before he was allowed to drive and steal other things from his family. He got into fights."

Mitchell says things escalated from there. The young man decided to break into people's homes to steal valuable items that he could sell.

"He broke into the first home, and there was nothing there," Mitchell recalls being told. "He broke into a second home, and there wasn't anything there as well. He was getting increasingly frustrated. He broke into a third home and found that someone was home sleeping. He just lost it and physically attacked the person."

The victim survived the assault but was traumatized severely. At the trial that led to the young man's conviction, several victim impact statements were read in court. Mitchell asked the young man how those statements affected him and was startled by his reply.

"He said to me, 'They keep talking about remorse and how it's affected me. What I don't understand is that I'm the one in jail.'" Mitchell looks straight ahead as he tells me, "He just couldn't grasp the concept of why the victim was traumatized and afraid to sleep in his own bed."

What surprised Mitchell even more was the young man's otherwise normal bearing and demeanour. "If you met him at the park and chatted with him, he'd be like anyone else. There was just this one bit that was missing and it really was foreign to him. That was a really defining moment for me. From that point, I thought that this isn't just about having a nasty environment. There's something fundamentally different about his thought processes. And I became really interested in looking at it from the perspective of the brain and neural functioning."

Mitchell was determined to learn the causes of psychopathy. He needed a mentor and found one in James Blair, a British neuropsychologist who has devoted much of his career to the study of psychopaths. Mitchell got a job working with Blair and moved to the Institute of Cognitive Neuroscience at University College in London, where he obtained his PhD. They continued their association after Blair moved to the United States to become chief of the Unit on Affective Cognitive Neuroscience in the Mood and Anxiety Disorders Program of the National Institute of Mental Health Intramural Research Program in Bethesda, Maryland.

At the time, Blair was moving away from the standard psychiatric definition of antisocial personality disorder, which focuses on observable behaviours, such as failure to obey laws and norms, lying, manipulation, irritability, irresponsibility, and lack of remorse. He was more interested in what he believed to be the root cause of psychopathy: a profound disturbance in responding to emotional cues from others, marked by reduced guilt and, most importantly, reduced empathy.

Mitchell did formative work with Blair mapping out the abnormal thought processes of psychopaths in the prison population. "We did some work in an open prison in Grendon in the countryside in England," Mitchell recalls. Grendon is a men's prison operated by Her Majesty's Prison Service and located in the village of Grendon Underwood, in Buckinghamshire. It operates as a "therapeutic community" that houses violent offenders and has far fewer restrictions than most prisons.

At the time Mitchell was doing his PhD, fMRI brain scans—similar to the one Philip Jackson took of my brain in Quebec City—were becoming an essential way to peer inside the brain. But accessing the technology to study Grendon's prisoners posed some unique challenges: The fMRI scanner was in London, some 110 kilometres away.

"We were always afraid they were going to escape," he says. "It was one thing to have an individual at the end of their sentence in an open prison in the countryside. Bringing them into London was another matter. We had to get guards to accompany them for that reason. Logistically it was very difficult, and it was always very stressful."

The lack of quick and easy access to brain imaging forced Mitchell and Blair to find an easier way to test prisoners. They hit upon neuropsychological assessment tools—paper-and-pencil and computer-based tests that are similar to the ones Mitchell had me take.

The time-honoured paradigm is that psychopaths are mean or cold-hearted, as evidenced by their inability to respond to the emotional distress of others. The tests administered by Blair and Mitchell and by others have pointed to a radically different explanation.

Consider that you are late leaving the house for work, and you run to catch the bus. Along the way, you see a child who has fallen off her tricycle and is crying. Your primary goal is to get to work on time. But your brain takes in the sound of the child crying and the look of distress around the child's eyes. In a split second, your brain processes those bits of what psychologists call secondary information and begins to shift your attention away from getting to work, compelling you to decide whether attending to the child is more important than getting to work on time.

The psychopath sees the same situation quite differently. The psychopath's brain is literally hijacked by the goal of getting to work. It uses so much brainpower paying attention to the primary goal of getting to work that it is unable to process secondary information from the child in distress. Neuroscientists call that inability an "attention bottleneck," a flaw in brain processing that makes psychopathy seem more like colour-blindness than cold-heartedness.

In addition, Blair and Mitchell found other flaws in the ways psychopaths process emotions. The researchers noted that psychopaths have profound difficulty recognizing facial expressions that show fear, sadness, or joy, yet have a normal ability to recognize faces that show anger or disgust. They also don't respond to the pain and distress experienced by others. Compared to normal people, psychopaths are less likely to stop doing something that causes others to feel pain or distress. Blair, Mitchell, and others began to formulate that these and other disturbances in thinking and feeling point to abnormalities in some specific parts of the psychopath's brain.

"We were looking at an overt behaviour that we could see from the results of tests with paper and pencil or on a computer," recalls Mitchell. "We had these hunches. We were making inferences about what bits of the brain are not functioning properly."

The "hunch," as Mitchell put it, was that the disturbances in thinking and feeling that are characteristic of psychopaths pointed to a problem in a part of the brain called the amygdala.

It's a pair of almond-shaped bundles of nervous tissue located deep inside the brain—one on each side. Neuroscientists say the amygdala's main jobs involve the processing of emotional reactions, memory, and decision-making.

The theory came first; the fMRI brain images that confirmed it would not come until much later.

Blair, Mitchell, and their colleague Karina Blair wrote about their findings in *The Psychopath: Emotion and the Brain*, a seminal book published in 2005. "We believe that the amygdala is functioning atypically from an early age in individuals with psychopathy," the authors wrote. "Furthermore, we believe that it is this problem in amygdala functioning that leads to the psychopathic individual's impairment in emotional learning. We believe that this impairment in emotional learning is at the root of psychopathy."

Since joining the Brain and Mind Institute at Western University, Mitchell has tried to figure out what specifically the amygdala does in relation to empathy. In 2011, the rising star and his colleagues compared research volunteers who were considered callous with volunteers considered kind. Each research subject had an fMRI brain scan while looking at a photo of a person showing an expression of fear.

Unlike the kind research subjects, the ones who tested as callous found it difficult to empathize with photos of people who had fearful facial expressions. And the fMRI brains of callous subjects showed far less activity in the amygdala than the kind subjects.

Bingo.

"When we finally did get around to imaging, the areas we had identified with the paper-and-pencil tests were the same areas that sophisticated imaging was identifying as functioning abnormally," says Mitchell with satisfaction. "It was really remarkable that we were able to figure much of it out without the technology."

Elsewhere, other top researchers have had similar results. In 2013, Jean Decety, the developmental neuroscientist and empathy guru who mentored Laval University's Philip Jackson, led a study of 121 inmates with varying degrees of psychopathy at a medium-security prison in the U.S. The inmates were shown close-up images depicting a painful mishap, such as a finger caught in a door and a toe caught under a heavy object. The inmates were asked to imagine being the person experiencing the mishap; they were also asked to imagine it happening to someone else. Their brains were scanned with fMRI as they viewed the images.

When those inmates with the highest scores for psychopathy imagined their own pain, the regions of the brain involved in empathy, including the right amygdala, anterior insula, anterior midcingulate cortex, and somatosensory cortex, lit up. If anything, psychopaths are more sensitive than non-psychopaths to the mere thought of being in pain *themselves*. But when the same inmates were asked to imagine the pain in others, the same regions failed to light up on their fMRI scans. Disturbingly, these psychopathic inmates showed *increased* activity in a region of the brain called the ventral striatum, an area known to be active when feeling pleasure. That finding raises the possibility that psychopaths enjoy imagining pain inflicted on others.

If there is a faint hope regarding treatment for psychopaths, it rests in the fact that their brains are not damaged; they're just different. Perhaps they can be trained. Contrast that with patients who have had a stroke and undergo personality changes, including loss of empathy. The stroke has damaged the amygdala in ways that cannot be repaired.

Mitchell takes me to a room down the hall from his office to witness a study being led by one of his postdoctoral fellows. A young man sits at a computer. He looks relaxed. A woman in her late twenties instructs the man.

"In this task, you will view a series of faces, one per trial," says Joana Barbosa Vieira, a psychologist doing research on empathy. He is asked to look at the faces and adjust the size of them until he feels he is at a comfortable distance from them. As a reference, he is asked to keep in mind the distance he would normally keep when having a conversation with a total stranger.

The young man begins looking at the faces. Some are men and some are women. They depict different emotions: joy, happiness, fear, anger, sadness, and surprise. Some have a neutral expression that reveals no emotion. In the photos, the eyes look left or right or straight ahead at the young man. Sometimes, the young man zooms in on the face, making it larger; other times, he zooms out to make the face smaller.

Barbosa Vieira explains the purpose behind the study. "We are looking at the distance that people choose to keep between themselves and others when they are interacting," she says.

She says that moving the face closer means the young man is approaching the person in the photo; moving it away means he is avoiding him or her. In this experiment, it's photos only. In other experiments, they use live actors to portray a particular emotion. That way, Barbosa Vieira explains, she can measure both the subject's intention to avoid or approach the actor and get the subject to predict whether the actor is likely to reciprocate.

"One of the things we are interested in is whether specific facial expressions are associated with motivation to approach or avoid the person," she says. "For an emotion like happiness, if you're the test subject, we predict that you'll think the actor is willing to approach you, and you will be willing to approach the actor. For emotions like anger, in theory, if someone is angry, they are going to approach us and we want to avoid them."

The experiment is designed to measure the subject's emotional empathy for the actor. To decide whether to approach or avoid the actor, you have to be able to identify correctly and respond emotionally to the facial expressions that you're seeing.

The main emotion Barbosa Vieira studies is fear. She wants to understand how good empathic test subjects are at responding to an actor who appears frightened. "If the actor looks afraid, this is a cue that there is something wrong or threatening in the environment and so it would prompt the test subject to escape," she says. "But fear may also be a cue that elicits empathy and a desire to help the person."

She says experiments like these are designed to be helpful in understanding empathy by mirroring real-life dilemmas and experiences. Surprisingly, the young researcher tells me she doesn't need to study empathy in criminals. "For a long time, it was understood that to study psychopathy you would have to go to an antisocial population, either a clinical population or a prison population," she says. "But we've gone a long way since then. We can assess those traits like callousness in a normal population."

Callousness is a clinical term that refers to things like a general disregard for others and a lack of empathy. Studies have shown that up to 25 percent of prisoners are psychopaths. In the non-prison population, just 3 percent of men and 1 percent of women are psychopaths. However, as many as 15 percent of the general population have callous traits without being diagnosed as psychopaths, according to the 2012 book *Almost a Psychopath: Do I (or Does Someone I Know) Have a Problem with Manipulation and Lack of Empathy?*

That's a much larger pool of subjects for researchers like Joana Barbosa Vieira and Derek Mitchell to study. "We look at community samples," she says. "We use self-report measures that do not rely on overt antisocial behavior but rather on how people score in specific personality traits like low empathy, for example."

And how do they tell callous people from the warm and fuzzy types? By using tests such as PPI-R, the same test that Mitchell had me take.

"There is a practical advantage of having access to bigger samples and less obstacles to go through, because you are not going into prisons and working with inmates," says Barbosa Vieira. "Simply put, you need to have a big enough sample that you ensure that you have a good range of psychopathy scores. So that's what we have been doing."

For her supervisor, Derek Mitchell, a big part of working out the basics of psychopathy is figuring out the balance between nature and nurture. Psychopathic traits tend to run in families, but upbringing is also important. In general, Mitchell says that a callous person who grows up in an affluent neighbourhood with parents who wield social influence is likely to end up far different from one who grows up in an impoverished environment.

Like Barbosa Vieira, Mitchell has shifted his attention from psychopaths in prison to people who have the traits but whose behaviour falls far short of the diagnosis. He has also shifted his focus from adults to children who may be psychopaths in the making. Developmental pediatricians refer to these children as having disruptive behavioural disorders, a group of conditions that includes oppositional defiant disorder (ODD) and conduct disorder (CD).

"In the case of ODD, they have a really severe problem dealing with authority," says Mitchell. "They're very difficult to manage for adults, parents, and teachers. They talk back, they can have temper tantrums when they're asked to do things, and they can become violent. Kids with conduct disorder (CD) often have similar symptoms and also engage in stealing, running away, precocious sexual activity, and the use of weapons. It's considered a more severe form of ODD."

Behind the labels is a fundamental difference in the behaviour of these children. Some do aggressive things without regret. Others do the same things but feel remorse afterwards. These kids, says Mitchell, might be treatable.

For Mitchell, that's the hope that makes his research so satisfying.

"These kids are still developing," he says. "Their brains are still developing, and if there's an opportunity to intervene and get these systems online and functioning, then childhood is the time to do it, as early as you can. We're really hopeful that we can train youth to activate the system that leads to empathy. If you're able to do that through training, then maybe you can have an impact on the way these systems develop and emerge in later adolescence and adulthood. Maybe we can make a difference for them."

Perhaps Mitchell finds it easier to empathize with children who have at least some hope of avoiding the outcome that awaits them. Perhaps he's empathizing with the parents of these children whose lives move inexorably from frustration to heartache. Perhaps he needs to focus on a population that is not beyond hope.

Whichever it is, Mitchell has found a way to study psychopathy by empathizing with psychopaths. Correctional and law enforcement officers, psychotherapists, and even some of the top researchers focus on antisocial behaviour, which, by definition, is awful. Who among us can empathize with a rapist or a murderer?

Here's what Mitchell has taught me so far. By focusing away from the behaviours and toward the cognitive and emotional processing that leads to those behaviours, Mitchell has managed to see these complex individuals as human beings. They're not evil, he might argue; they just harbour a disorder of neural processing much like the diabetic who lacks insulin. In this way, Mitchell takes away the fear of being manipulated and even hurt by psychopaths. Take fear out of the equation, and it's even possible to empathize with them.

ER physicians like me can learn a lot from Mitchell. We define patients with outsized undesirable traits by their behaviour—the agitated patient who can't control the impulse to scream or punch his fist into a wall in the psychiatric assessment room. That Mitchell has helped me see them in a different way is nothing short of remarkable.

Derek Mitchell has shifted his focus from hardened criminals to everyday people who possess some psychopathic traits yet aren't twisted enough to land behind bars. At Emory University in Atlanta, another top researcher named Scott Lilienfeld has made his career examining the behaviour of a different kind of psychopath. This kind manages to stay out of prison and often generates a surprising amount of admiration.

Psychologists like Lilienfeld call these people successful psychopaths. At the top of the list is a man named Forest Yeo-Thomas, a British spy who was almost certainly the real-life inspiration for Ian Fleming's agent 007, James Bond. "We wanted something that was kind of catchy," Lilienfeld tells me in an interview. "Given almost everyone has heard about James Bond, we thought that would be a good one to start with."

Lilienfeld wrote the book on successful psychopaths, whom he describes as "someone who can be quite ruthless, self-centred and mean, but by the same token also comes off as very charming, appealing, and quite alluring to people who don't know him well." Examples abound, but Lilienfeld says that successful psychopaths tend to gravitate to occupations where their flamboyance attracts fans. The list includes CEOs, politicians, religious figures, scientists, lawyers, physicians, and symphony conductors.

Lilienfeld recounts the true story of a conductor of an orchestra in North America. Much of what he tells me comes from a musician who played in the orchestra for years during the conductor's tenure. "The conductor was very socially dominant and was much beloved by benefactors," says Lilienfeld. "This person was very good at raising money for the orchestra and would charm the pants off of lots of people in the city in which he lived." His musicians, however, saw a different side to the conductor: a bully to some, a monster to others. "He would delight in humiliating certain performers during practice sessions," says the psychologist. "He could be quite mean and vindictive. He had a long history of philandering, lying, cheating on his wife, and stealing."

As often happens with successful psychopaths, those who observed or suffered his bad behaviour could not convince those who admired him. Lilienfeld says they are crafty at nurturing the support of influential people by never showing their bad side to them.

Unlike the psychopaths that Derek Mitchell studies, the successful psychopaths studied by Lilienfeld are able to avoid behaviour that is overtly criminal. That was true in the case of the orchestra conductor. "He was never physically violent," says Lilienfeld. "He never broke the law in any blatant way that would put him in prison. That's the prototype we're talking about."

There's something about the way Lilienfeld describes the conductor that makes him seem like a narcissist. The psychologist says successful psychopaths possess a lot of narcissistic qualities. There's so much overlap between the two traits that he thinks it's pointless to say that someone is one or the other.

Lilienfeld says that successful psychopaths tend to be what psychologists call grandiose narcissists. Like the conductor, they have an overwhelming need for admiration and seldom appear vulnerable to others. They tend to be emotionally cold and rarely if ever empathize with people, even admirers. They are full of pride and easily angered.

Another type of narcissist is the vulnerable narcissist. Lilienfeld says people with vulnerable narcissism have low self-esteem. They are preoccupied with the fear of being rejected and are very defensive when confronted about mistakes and foibles. When I ask if he can give me an example, Lilienfeld immediately thinks of famous political figures.

"The vulnerable narcissist might be more like a Nixon or someone like that, just kind of thin-skinned," he says. "The grandiose narcissist might be more like a Trump, or a Mussolini, for example. People like that can be quite appealing to some people. Some people may find them abrasive, but others might like them because they are kind of flamboyant and entertaining. They often have a sense of humour. Often, they are freed from the typical social inhibitions that most of us have, so they can appear quite charming."

Lilienfeld says that successful psychopaths are also risk takers. Their ranks include daredevil sports figures, law enforcement officers, firefighters, paramedics, and career military officers. What they have in common is something called fearless dominance.

"We think that fearless dominance reflects things like physical and social boldness, resilience to stressors, poise, confidence, and charm," says Lilienfeld. "I and others think that their brains are less sensitive to threat and less sensitive to danger. In some cases, that can be dangerous because people may say and do stupid things from time to time. By the same token it can also be advantageous because these are people who are more adventuresome, and more exciting, and may be more attractive to potential romantic partners."

It makes me wonder if paramedics and firefighters would be more predisposed to fearless dominance than other health professionals, and perhaps more likely to be successful psychopaths. "They may be more likely to have fearless dominance," Lilienfeld says. "I'm not sure they're more likely to have the other traits of psychopathy."

Lilienfeld says both narcissism and psychopathy are on the rise. One reason may be the increasingly rapid pace of society. "It could be that we live in a society where people can advance a bit more quickly," he says. "There's more social mobility than there once was, which in general is a good thing. But that may also enable people who are unscrupulous or ruthless to take advantage of others and rise more quickly."

Other aspects of modern life may set the table for successful psychopaths to flourish. The growth of social media makes it easier for the successful psychopath to charm larger numbers of people than would have been possible decades ago.

But what stops them from being criminals? Neuroscientists have several theories. For one thing, criminal psychopaths tend to have slow pulses and slow breathing as they contemplate and commit crimes. When successful psychopaths contemplate criminal acts, their heart rate and breathing speed up, perhaps making them uncomfortable enough not to commit the crime. Compared to criminals, successful psychopaths also tend to have higher executive function—the mental capacity that organizes our efforts to get things done. That gives them the mental flexibility to channel their wants and desires into activities that aren't crimes.

Lilienfeld's stories about nasty conductors and politicians are entertaining, but I'm writing a book about kindness and empathy. Does the rise in successful psychopaths mean that empathy is going down? His answer is sobering.

"Empathy is a trait that can be used for either good or bad purposes," he says.

Lilienfeld is talking about cognitive empathy, which I defined earlier as the capacity to understand another person's thought processes and predict what they are likely to do in certain situations. Studies show that psychopaths, especially successful psychopaths, are very good at cognitive empathy. "If you're really good at reading other people and predicting how they're going to react to you, that is a good trait," he says. "But if you are someone who is already unscrupulous and cold-hearted to begin with, it is also a trait that can be misused and maybe even used for evil if one isn't careful."

I'm certainly no James Bond. From the moment Lilienfeld started talking about the orchestra conductor, I knew I wasn't a psychopath, successful or otherwise. I'm way too insecure to be considered grandiose. But vulnerable narcissist is not nearly as big a stretch. For most of my adult life, I've been a man with somewhat fragile self-esteem. Vulnerable narcissists are prone to anxiety and depression, as am I. When criticized, they feel shamed and humiliated rather easily. That has certainly been one of my core issues over the years.

There is an intriguing connection between empathy and narcissism. Studies have shown that people with narcissistic personality disorder (NPD) have impaired emotional empathy, but cognitive empathy may be largely intact.

This is where the distinction between grandiose and vulnerable narcissism becomes important. Recent studies have suggested that people with grandiose narcissism can dial up or down cognitive empathy depending on what's in it for them. Vulnerable narcissists are different. If anything, they have an exceptional capacity for empathy, so much so that friends and family think they have emotional antennae. What they don't do well is take criticism. As long as they are safe from criticism, they feel okay. The moment they are criticized, the resulting emotional turmoil impairs their ability to empathize with others, making them withdraw.

Much of that sounds like me. The thing is, it also sounds like a lot of health professionals I've known and worked with over the years.

I reached out to Lilienfeld thinking I'd learn a lot about swashbucklers and CEOs. I've ended up learning some surprising things about myself.

* * *

I'm holding my breath as Derek Mitchell, the Western University neuroscientist, is ready to give me the results of the tests he administered. I feel like I've been called into the doctor's office to get the results of a biopsy.

"One of the things we looked at is the extent to which you show some of the traits associated with the autism spectrum," says Mitchell, referring to the Autism-Spectrum Quotient (AQ), the questionnaire developed by British neuroscientist Simon Baron-Cohen. Mitchell chose that particular test because it measures cognitive empathy, something people with autism typically lack. The AQ also looks for other symptoms of autism, such as excessive attention to detail and difficulties relating to others.

"You're not high on the autistic spectrum," says Mitchell, eyeing the test results. "You score pretty much bang in the middle for males, who tend to score slightly higher than females. Individuals who have careers in mathematics and the sciences tend to score higher than individuals who are maybe more on the artsy side of things. You're bang in the middle there too."

He's telling me I don't have autism, which is hardly a shock. The next test result is my PPI-R, the psychopathy personality inventory.

"This one is interesting because it measures your psychopathic traits, but it's broken down into a number of things," says Mitchell. "You'll be happy to hear that you're not high overall in psychopathy."

Relieved is much closer than happiness to what I'm feeling. Then, he starts to give me some numbers.

"In fact, one of the things we're really interested in at our lab is cold-heartedness," he says. "As the name implies, cold-heartedness refers to individuals who are low on empathy.

They are not moved emotionally by other people. You score at the 15th percentile, so you're very low for cold-heartedness, which is good. I talk about people scoring like that as the 'cuddly' types, so there you go."

I'm not just low in cold-heartedness, I'm cuddly! For a guy who went looking for his lost capacity for empathy, I've hit the jackpot. But Mitchell is not finished giving me his findings.

"There are other indices that are higher, which is interesting." He emphasizes the word *interesting*. "Your score for Machiavellian egocentricity is at the 86th percentile," says Mitchell.

For a moment, I'm speechless.

Niccolò Machiavelli is the Italian Renaissance writer and political theorist who penned the sixteenth-century treatise *Il Principe* (*The Prince*). In the book, Machiavelli states that the goals and aims of the prince justify the use of immoral acts to achieve them.

I ask Mitchell to elaborate. He chooses his words carefully.

"Machiavellian refers to—I don't want to say 'using' people, but being able to get at what you want, and to get that out of other people as well," he says. "It also refers to a tendency to possibly be a bit more cynical about others, about human nature, and to conduct oneself accordingly."

The PPI-R questionnaire defines *Machiavellian egocentricity* as "a tendency to consider only personal needs, often disregarding the interests or perspective of other people." It also refers to something called *social potency*, a term that means "the ability to charm and influence others."

"Have I got that too?" I ask Mitchell.

"Your social influence is also at the 86th percentile, so you're high on that element too," he says. "That would correspond to the ability to get people to do what you want them to do. You view yourself as capable of that and you're probably adept at it."

The third thing that the PPI-R measures is *impulsive nonconformity*, defined as "disregard for social norms and culturally acceptable behaviours."

"You're at the 71st percentile for rebelliousness and nonconformity," he says. "So, you're a more free-spirited rather than by-the-book sort of individual."

As I'm thinking about all of this, I notice that Mitchell's demeanour has changed from affable to serious. I ask him how he's feeling.

"I never go through this with our participants," he says. "We don't go into that level of detail. Usually, they are not interested. I don't have trouble discussing the social influence part, but I'm really uncomfortable talking about Machiavellian egocentricity. I think part of it is because it's less easily described in a flattering light."

Mitchell adds that there's another reason why he's being circumspect in drawing firm conclusions from my test results. "I don't want to give you the wrong impression," he says. "I don't want to imply that anything is pathological here."

By pathological, Mitchell is referring to the dark triad of personality disorders that Paulhus and Williams first described in 2002. Mitchell says the PPI-R he had me fill out is not intended to diagnose a personality disorder; it is designed to detect personality traits in *normal* people. "This is a tool used in a community sample and this is the normal range," he says, looking at my test scores. "That is part of the reason why I was hedging it. None of these numbers are a concern."

I decide to address the elephant in the room. "You want to ask me if I was surprised by the results."

"Sure," says Mitchell.

My first reaction to Mitchell's assessment *was* surprise. The second was embarrassment at having the scientist know that I possess some attributes that are, shall we say, less than flattering. My third reaction is one of grudging acceptance.

"I guess I'm not surprised about the test," I tell him. "I think that's a part of what makes me a decent medical journalist."

I started this journey by admitting that I'm not always the kind and caring physician I aspire to be. But medicine is not my only profession. I've been a journalist for more than three decades. I've spent the last 10 years hosting *White Coat, Black Art*, a CBC radio documentary series that examines the culture of medicine. As an interviewer, my stock-in-trade is getting people to reveal their innermost thoughts. I'm good at it or I wouldn't have a career in journalism.

"Some degree of social influence and ability—how shall I put this—to interact with people in ways that get them to do things that are useful to you can be beneficial in a lot of areas," says Mitchell.

He's saying I'm good at persuading people to be interviewed, an essential part of my radio job. What's true in journalism is even more so in politics.

"I think politics depends to a large extent on being able to influence people and figuring out what it is they might want in order to push your agenda," says Mitchell. "To get to a position where you're having influence, you need those traits."

When I conduct an interview, I use my intuition and an exceptional ability to read emotions to connect with the guest. There are times when they end up saying things on tape that I'm sure they regret. Secret things. Intimate things. Things that make the listening audience sit up and take notice. When that happens, I feel great satisfaction that I've done my job exceptionally well. But what about the person I've interviewed? Having been interviewed myself, I know there are times when they say things they wish they had left unsaid.

Knowing that I have a Machiavellian side makes me wonder just how often it plays a role in my success. Have I used my powers of persuasion to get people to answer questions they would rather avoid? How often have I put my ambition ahead of their comfort and dignity?

If this were the medical world, and the interviewees were my patients, my duty not to harm would be paramount. I would protect them from their worst impulses to reveal themselves. In the world of journalism, my duties to my interview guests are balanced by my duties to the broadcaster and the audience.

"I think that's why some people might be reluctant to talk to the media—because they feel there's an agenda," Mitchell says. "Depending on the individual, you might be very up front about what that agenda is or you might spin it, right?"

Unfortunately, Mitchell is right. While journalists should not lie to interview guests, they may not spell out in detail the potential fallout of being interviewed in the media. It makes me wonder if my Machiavellian side has played a role in the instances I've been guilty of this.

Now, it's Mitchell who notices that I'm uncomfortable about the implications of what we've been talking about. He reminds me that I also scored high for empathy.

"Think about it as a combination of traits," he says. "If someone is relatively empathic and a decent human being, then scoring high on some of these indices such as social influence is probably not that scary a thing. However, if we were to combine your high social influence scores with high cold-heartedness, then it would be another matter altogether."

Mitchell tries to reassure me that my "cuddliness" balances out my Machiavellian tendencies. But I know now that there are times when my desire for fame and accolades overrides my ability to empathize with others.

On my way home from Derek Mitchell's lab, I am disturbed by what the testing has revealed about my character. I find myself ruminating over the findings. I imagine every health professional I've ever interviewed accusing me of burning them. I find myself mentally agreeing with each and every one of them.

My skin feels hot as if I have a fever. I feel heavy in my chest and sick to the point of nausea in my gut. I feel exhausted. I feel unworthy of love and friendship. I find myself wishing I could disappear.

When I get home, the symptoms begin to abate. I know I'm not having a heart attack, but I google the symptoms nonetheless. I get lots of hits for anxiety and panic attacks. But one makes me stop. It says that the symptoms I'm describing are typical of people who feel shame. Shame about being told by Mitchell that I have Machiavellian tendencies, even though he spoke to me with great kindness. Shame about the people I may have harmed because of that tendency. Shame about the times I haven't been kind to patients and families.

So far, I've had my brain scanned and my personality assessed. It's all been about me. It's time to go out into the real world and see what I can learn from people who are known for their kindness to others.

CHAPTER FOUR

The Donut Shop

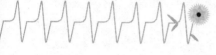

It may seem strange to search for empathy in a donut shop—
especially on a cold and blustery day in March—but I've got a
line on a man people say is one of the most caring guys in Canada.

I drive along Highway 401, the main east-west highway cutting through Toronto, to a Tim Hortons on the northeast corner
of Ellesmere and Neilson Roads. The restaurant sits in the heart
of the eastern suburb of Scarborough, just a few kilometres
from the Scarborough Bluffs, a handsome escarpment along the
Lake Ontario shore. My parents used to take me there when I
was a kid. A chill wind is blowing north from the bluffs. I pull
the collar of my winter coat to my neck as I walk into the shop.
The freestanding structure in the parking lot of a strip mall is
actually two fast food joints—Tim's on the right, Wendy's on the
left—with a common entrance under a sharply angled roof.

The pastel-coloured interior is prefab but spotless. The
decor says: "'Stay awhile, but then please move on." The lineup
of customers is 20 deep, but it's moving very fast. Three people
behind the cash are taking orders, pouring coffee, and serving
donuts; another four are preparing hot food. It looks like any

other Tim Hortons—one of 4,590 around the world and 3,665 in Canada—a chain that boasts on its website that it serves a staggering 2 billion cups of coffee a year. This location serves 2,500 customers a day, a very large number in a highly saturated and ultra-competitive market.

The tables are crammed. A 20-something man in a three-piece suit sits alone and taps on his notebook computer. Four teenage girls dressed in grey private school tunics are drinking Iced Capps. They don't talk, just text on smartphones while giggling periodically, together yet separate. Next to them, four helmeted construction workers in full safety gear wolf down breakfast sandwiches. Near the window, eight retirees chat at two large tables pushed together. A slightly overweight middle-aged man with the face of a 10-year-old boy moves effortlessly from table to table—smiling and joking with each customer. Young, old, it doesn't matter. Each person stops to shoot the breeze with him. Even the giggly girls in tunics look up briefly from their smartphones to smile and say hi.

Who holds court at a Tim's? I'm curious, but he's not the guy I'm here to meet. He's late. To kill time, I walk up to the counter and order a small dark roast with half cream. I start to hand over a toonie to the young woman at the cash. Suddenly, a large fist propelled by a muscular forearm bumps into my hand. Slightly annoyed, I look up and see a man in his mid-fifties grinning at me from the business side of the counter. "I'm Mark Wafer," says the man, laughing. "At my restaurant, the coffee is on me."

I begin to take fuller measure of my host. Mark Wafer is a bit taller than I and powerfully built, with a thick torso and neck, thinning grey hair, and a white goatee that is groomed neatly. A boxer and a businessman rolled into one. Don't know about you, but I've never heard anyone call a donut shop a restaurant. It's the first of many indications that Mark takes his business very seriously.

Mark invites me to his side of the counter. I find myself smiling. I'm getting the special tour of the kind of place I've frequented thousands of times.

Mark Wafer owns six Tim Hortons restaurants that serve 13,000 customers a day. Each is located in a busy part of Scarborough, each among the most profitable stores across the chain. Each, in its own way, a smashing success.

Mark takes me to each of his stores in his pickup truck. The first stop is a drive-through-only location on the corner of Markham Road and Lawrence Avenue, steps from a major urban transit hub. No tables and chairs; just walk up or drive through. On the way into the store, Wafer shows me the exterior feature that doubles the profit. "It's what we call a double-drive-through location," says Wafer. Cars enter one of two parallel order-entry lanes; they merge into one lane in which drivers pick up their orders and pay for them. "It means we get twice as many cars in," says Wafer, his blue eyes twinkling. "There are no seats and no bathroom. It's what I like to call a profit centre."

The key to success in a donut shop like this is high volume and fast service. On the inside, I count eight uniformed employees processing orders or handing them to drivers as they pass through, a beehive crammed into a tiny space. This store serves twice as many drivers as others with a single drive-through lane, and the servers are very fast at processing orders.

"What was your best window time today?" Mark asks Gracia, the manager of the location.

"Nineteen seconds," says Gracia.

"What time was that at?" asks Mark.

"Six to seven A.M." she replies.

"That means from 6 A.M. to 7 A.M. this morning, the average time of every single car sitting at that window was 19 seconds," says Mark with a smile. And that's not even the fastest time; Mark says at another location, the average wait time for drivers at the window is as little as 14 seconds.

No one is stopping to appreciate the compliment. The workers move with blazing speed, some taking food off the shelves, some making sandwiches and salads, and others constantly bringing baked goods just out of the oven from the adjoining kitchen.

Though it's orchestrated chaos, the vibe is stress-free. It feels like a happy place.

Mark has shown me the importance of speed and productivity. He takes me back to the store where we first met, to show me the real secret of his success. We go past the counter into the kitchen, where he points to a blonde woman in her mid-thirties named Jennifer McCall. She started working for Mark as a 15-year-old. That was 21 years ago. McCall got promoted early and often and now is part of Mark's management team. She's talking to the middle-aged man I saw greeting customers earlier that morning. Only now, I realize that he's wearing a Tim's uniform.

"Brian, I'd like you to meet Clint Sparling," says Mark. "Clint has been working at Tim's for over 20 years."

Clint Sparling, now 44, is one of the first employees Mark hired when he bought his first Tim Hortons in the mid-1990s. He's worked for Mark ever since. Clint extends a fist for me to bump. He has a small upturned nose, large, rectangular-shaped glasses, round eyes that are dark brown and kind, and a big, open-mouthed mischievous grin.

Clint has Down syndrome.

* * *

In the early 1990s, Mark Wafer and his wife, Val, decided they wanted to buy a business. Until that point in his career, Mark's biggest success had been as the operations manager for a major car dealership. He was instrumental in making the service department the most profitable part of the business. The next step seemed logical.

"At first, I thought it might be a car dealership," recalls Mark. "We were given the go-ahead by a car manufacturer to become a franchisee, but at that time the amount of capital required would have meant having financial partners."

At the time, Val worked as an accountant at the corporate head office of Tim Hortons. "With a Tim Hortons, you don't need partners," says Mark. "My wife and I would be partners."

Today, the fast food market is saturated with Tim Hortons and competitors. Back in 1995, the chain had few locations in Toronto and surrounding areas. There was one franchise in Scarborough, and the owner wanted to sell. "We knew if we bought one store, then other stores would come along very quickly because there was nothing here at the time," he says.

Mark and Val bought the franchise in September 1995. It was a turnkey operation, complete with equipment, employees, and a lot of customers. The first week they owned the store was an eye-opener. "My staff couldn't keep up with the dining room, the tables, the dishes, and so on," Mark recalls. "I had a need to hire somebody. I put an ad on the board: 'Dining Room Attendant Needed.' And in walked Clint."

I ask Clint what he thought about Mark at that time.

"Well, at first I didn't even know Mark that well," Clint recalls. "I went over there and I told Mark I was a clean freak and so he signed me up."

Clint was 23 years old. He had done some volunteering. That he hadn't worked a day in his life didn't matter to Mark.

"When Clint walked through that door, I knew I was going to hire this guy," he says. "He hadn't even got two words out of his mouth. I knew I was going to hire him because there was no way he was going to find a job anywhere else. There was no way that any employer was going to hire him, so I gave him a chance."

Deciding to hire Clint was the easy part. Figuring out what Clint could do and how to train him to do it was more challenging.

Down syndrome is a genetic condition caused by an abnormal chromosome. According to government figures, in Canada prevalence of Down syndrome averaged 15.8 per 10,000 total births between 2005 and 2013. Down syndrome causes short stature and characteristic facial features, plus heart, bowel, and other conditions. It also causes profound intellectual deficits.

Gainful employment among people with Down syndrome is far more the exception than the rule. In the United Kingdom, research from the charity Mencap suggests that fewer than 20 percent of working-age people with Down syndrome hold jobs. In the United States, a survey of nearly 500 adults with Down syndrome found that 25.8 percent were volunteering, and 30.2 percent were unemployed. Twenty-three percent had never had a job.

That's data from 2015, a time when legislation and advocacy for people with disabilities were in full force. Mark hired Clint in 1995, when the climate for giving people with Down syndrome a shot was much chillier.

"Everybody in the restaurant was looking at me and asking, 'Are you crazy?'" Mark recalls of the moment he decided to give Clint a shot. "I didn't know how it was going to work out. I didn't even know how I was going to do it."

Mark figured it would take longer to train Clint than other hires. And he didn't have the time to train his new employee and run the donut shop 24 hours a day, seven days a week. "I contacted Community Living Toronto. I knew that they had job coaches and job developers and I called them to ask for help."

Since 1948, the non-profit Community Living Toronto has supported thousands of youth and adults with intellectual disabilities who are looking for work. They do employability assessments as well as individualized planning and skills development.

"First, I was working on about how to learn everything about, like, going to work five days a week," says Clint, who worked part-time at Tim Hortons while being taught how to bring dishes to the dishwasher, clean tables, and sweep the floor. Community Living taught Clint things like how to take the bus to and from work.

Clint remembers with pride the day he finished the first leg of his training at Community Living Toronto for two reasons: Mark came to his graduation and he offered him a job.

Once employed, Clint graduated from basic jobs to dishwasher maintenance, a careful process that involves handling dangerous chemicals. From there, he began stocking items. Then he was mopping the floor, which is a skilled job in a restaurant because of the risk of a customer falling. Clint became proficient at floors and moved on to cleaning toilets. Slowly, his job description began to match that of many other employees.

There were things about Clint that got Mark's attention.

"He is so proud of his uniform," says Mark. "He used to come to work with his uniform military crisp, his shirt ironed, pants pressed, his hair perfect. He would spend five minutes putting his visor on because he wanted to look great. Back then, we didn't let our staff wear uniforms to work because they can get contaminated by the environment. Clint would wear it on the bus to let everybody know that he works at Tim Hortons. So, we made an exception in his case."

Jen McCall, the woman who was talking to Clint when we first met, is Clint's current supervisor. "He's just a regular employee to me," says Jen. "There's no difference between Clint and everybody else."

That doesn't mean he's perfect. Clint is his own worst critic.

"Once when I was working, I was cleaning this table," Clint recalls. "There was a bagel with peanut butter on it. I took it, wrapped it up, and put it in the freezer to eat later. Except that I got caught. Jen saw me and made me throw it out."

Mark treats Clint like any other employee. When Clint arrived late for work on one occasion, Mark docked his pay. Clint has never been late since. "Jen and Mark know what's best for me," says Clint. "They're just looking after my best interests and they know what's right from wrong."

A bond has developed between Clint, Mark, and Jen.

"We've become such great friends," says Jen. "Clint is like family to me. I'm not just his boss. I talk with him every day. I'm here with him almost every day. If he has issues, he talks to me, and sometimes I talk to his mom."

Mark says that Clint went from special project to coveted employee. "We saw a loyalty from Clint that we just didn't get from other employees," says Mark. "Halfway through 1996, when Clint had been with us for about six to eight months, we realized that Clint had become our best employee. He never came to work late. It was very hard to get Clint to take a break, and he didn't want to go home at the end of the day because the job meant so much to him."

Loyalty isn't just a positive attribute. As Mark discovered, it's a financial asset.

"When you hire somebody with all the expense of uniforms and training, you need a return on that investment," he says. "With Clint, the return is massive. He brings such value to this business. Not just in the job that he does but in the attitude he shows in doing his job. Customers love it. Customers come here because we hire people with disabilities like Clint."

The regular paycheque has enabled Clint to do so much more with his life. He summoned the courage to ask out a young woman named Katie that he had met at a swimming meet put on by the Special Olympics. In 2007, Clint and Katie got married and bought a condo. Clint has also become a motivational speaker.

"You look at the confidence he has today," says Mark. "This is outstanding. Being able to stand in front of a crowd of up to 200 people and tell them about his life takes a lot of confidence. He gained that simply by being an employee, getting a paycheque, and living a full life."

Without a paying job, Mark believes that Clint's life would be very different. "He would be making 40 cents an hour in a sheltered workshop," says Mark. "He wouldn't be married. He wouldn't be living in a condo that he and Katie bought. He wouldn't have a full life. He would probably be 40 pounds heavier and have depression."

What Mark Wafer got in return was satisfaction at Clint's growth and development, and a hypothesis about the economic value of taking people with disabilities out of sheltered workshops and parents' basements and putting them into meaningful jobs for decent pay.

The compassion Mark and Jen extended toward Clint—and many others, as you'll soon find out—is extraordinary. In the need-for-speed, cost-contained world of health care, that sort of kindness almost never happens.

It takes a huge wellspring of compassion for a fabulously successful entrepreneur like Mark to see the world from the perspective of someone with Down syndrome like Clint, or for that matter, a person with any other enormous challenge, be it physical, mental, or emotional.

Actually, it's not as difficult as you might think. Mark has a disability too.

Mark is deaf. He was born with 20 percent hearing in both ears and has even less hearing today. He has what doctors call sensorineural hearing loss, which means the nerves that turn vibrations into electrical impulses don't work properly. The Canadian Hearing Society says roughly four in 1,000 Canadian babies are born with some degree of hearing loss or develop a progressive form of it early in childhood.

While Mark is very good at lip-reading, phone conversations are a no-go. In every other way, he is completely able bodied.

Mark was born in London, England, in 1962. If he were born today, he would be a good candidate for a cochlear implant, a microphone and electronic device surgically implanted behind the ear that enables deaf people to hear by receiving sounds and sending them directly to the cochlear nerve. But cochlear implants weren't available until 1982.

Today, the United Kingdom, the United States, and six Canadian provinces have universal programs to screen infants for deafness at birth. But back in the 1960s, without universal screening, it was up to Mark's parents, Patrick and Rose, and his teachers to notice he had difficulty hearing. His parents were recent economic migrants from Ireland who met and married while driving double-decker buses in London.

"Looking back, my mother would say that I'd be in the backyard and she'd be calling everyone for dinner, and I took my sweet time coming in," jokes Mark. "The other brothers would run and I guess probably that was my cue, that they were running in so maybe I should go too."

As Mark recalls, his parents were preoccupied with making ends meet while taking care of a growing family. His mother had health problems of her own. "My mother and father didn't really notice my hearing," says Mark. "They were busy with three children 12 months and older, dad always working, and mum being sick. She had bronchial asthma, and it was the worst type. She died when she was 49."

In those early formative years, Mark thought he was like everyone else. Then he started school in a small community called Watford, northwest of London (a town later made famous when singer Elton John bought the local soccer team). That's where he was first treated differently.

"Growing up in Watford, soccer was huge," says Mark. "Every kid played soccer. In grade one, I remember the teacher taking me aside and saying I shouldn't play because I was going to get hurt." For Mark, who at the time was a good soccer player, the comment left him demoralized. "I was small in stature for my age and so I was intimidated by adults or older people who put barriers in front of me," he says. "I went along with it, but it had an effect on my self-confidence."

Back then teachers made snap judgments about the talents and abilities of students, especially those who had disabilities such as deafness. "I have a report card from when I was seven," says Mark. "Under the heading Music, it says 'Mark has no musical abilities.' I would be in the music class and everybody would be playing the clarinet or the recorder. I couldn't hear any of that and I'd get failed."

There was little understanding of accommodations or adjustments. The same attitudes prevailed in Ontario, where Mark and his family moved in 1974. Hearing aids were one form of accommodation that was permitted. Unfortunately for Mark, with his kind of hearing loss, listening devices made sounds louder but not clearer. "I kept telling everybody that they don't work, but my parents and my teachers felt that that was just me rebelling," Mark recalls.

He was damned for not wearing them, and on one memorable occasion, he was damned for putting them on. Mark recalls attending an assembly at which the principal spoke. Mark wore his hearing aids and sat in the front so he could read the principal's lips. In the middle of his address, the principal took notice of Mark and asked him to come up on stage immediately. "When I speak, I don't want to see any student listening to a radio," said the principal to Mark.

"It's not a radio," said Mark to the principal. "It's a hearing aid. I'm deaf."

The school had 1,800 students. "Everybody laughed," Mark recalls.

Through the eighth grade, above-average intelligence got Mark a B-plus average. Grade 9 brought larger class sizes, audio cassettes of Shakespeare that Mark couldn't make out, and trouble keeping up with French. He was close to failing. The principal noted his deteriorating marks, and Mark was transferred to a school for the deaf in Milton, Ontario, far from his home.

One week in, Mark found himself alongside students with severe intellectual disabilities. He told his mother he'd never graduate if he stayed at that school. "Mom pulled me out after the first week and I went back to the other school," says Mark. "I decided I was going to make this work. For the rest of my time in high school I had a tutor from the Canadian Hearing Society who helped me get through high school. And I did it—I got through and I went to college."

The lesson to Mark was that special schools are a route to being shunted from mainstream society. The fix is to stay in the mainstream by appearing as normal as possible, and by resorting to as few accommodations as one can. That's when Mark got proficient at lip-reading and tossed the hearing aids aside.

It's not hard to see the connection between Mark's experiences and normalizing the lives of people like Clint Sparling by giving them meaningful work and a paycheque.

But Mark learned even more hard lessons as he grew up. He says he endured taunting and bullying from classmates. Three teenage boys were the worst offenders. "They would chase me, and they would beat me, and they would throw me over a fence," he recalls. "It was all based on the fact that I was different, that I had a disability."

Later on, he encountered authority figures who didn't realize he was deaf. "The worst thing that can happen to a deaf person is to get caught speeding at night," says Mark. "When the police officer approaches your window, it doesn't matter what you say. You tell him you're deaf, he says you're being a smart aleck. You have a deaf card, but he doesn't want to see that. He wants to see your licence. He's asking questions, and you have no idea what he's saying."

Mark remembers getting pulled over by a police constable for driving with an expired licence plate sticker. It was August

1986, and Mark was 24. He says it happened early one morning as he drove to work. "He told me he didn't like my attitude," Mark recalls. "He told me to get out of the car."

He didn't understand what the constable was saying and began to protest. "He put me in handcuffs and took me to the police station. I had to sit in a jail cell for three hours to teach me a lesson. I didn't fight back, and I didn't retaliate."

Mark says that police officers are more respectful of deaf people now than they were 30 years ago. But Mark is also different. He vows that "today I'd definitely go to the Ontario Human Rights Commission."

A core of defiance and resilience that is apparent today was gestating back then. It was at this time that Mark also began to show flashes of the ability to empathize with people who have disabilities, which really came to the fore when he hired Clint Sparling.

When Mark was the editor of a student newspaper, he wrote stories on what it was like to be deaf and how important it was to give someone who is disabled a chance to succeed. That empathic streak temporarily led Mark down a very different career path to the one he is on today. The donut shop owner completed a diploma in social work at Sheridan College in Oakville, Ontario. He got a job at what is known today as Baycrest Health Sciences, a facility founded in 1918 as the Jewish Home for the Aged. Baycrest has long been a hospital, a long-term care facility, and a research centre that focuses on brain health and aging.

"I was a social worker for people who had Alzheimer's," says Mark. "I really enjoyed it."

But Mark had other plans. He left Baycrest after 18 months, because he figured he'd never make much money as a social worker. He watched friends who didn't go to college make a better living in the world of business. He switched careers, thanks to a penchant for taking risks and a passion for cars.

Mark Wafer has been racing cars since he was 22.

"I started out by driving in Formula Fords and then Formula 2000, which is an open-cockpit race car, similar to an Indy car but much smaller and much slower," says Mark. He now drives a 1988 Porsche 944 Turbo Cup, one of a few dozen factory-built race cars for North American tracks. Once driven by Canadian legends like Scott Goodyear and Paul Tracy, these vintage cars are now driven on the historic motorsport circuit by hobbyist drivers like Mark. They race cars from a particular era—but with modern safety precautions. Mark says he has raced on tracks all over the northeast, including Mosport Park in Bowmanville, Ontario, and Watkins Glen in upstate New York.

"Watkins Glen is the fastest track I've ever driven on," says Mark. "My car would do about 195 miles an hour on the straightaway. That's 315 kilometres an hour. Very fast."

Mark embarked on that dangerous hobby despite having a serious car accident on the highway at the age of 18. A tractor-trailer and Mark's car collided, and Mark's car was sent hurtling along the road. "We went into a gas station and took out all the pumps and so the fuel came up and I was burned by that. I was in pretty bad shape."

Mark broke his pelvis and spine in three places. He underwent several operations and spent nine months in hospital. By coincidence, Tom Wright, the orthopedic surgeon at Toronto Western Hospital who put Mark back together, just happened to be a rally driver. "He would come into my room and see me reading racing magazines," says Mark. "One day he says to me,

'I spent a lot of time putting you back together. You're going to promise two things. One, you're never going to jump out of an airplane and two, you're never going to drive a race car.'"

"I kept one of those promises," says Mark. "To this day I've never jumped out of an aircraft."

What drives a deaf man to want to drive a car 315 kilometres an hour? The smell of oil and gasoline and the engine vibrations are only part of why he does it. The other reason is a rebellious streak. "When somebody said I can't do something, even if I didn't want to do it, I was going to do it to prove them wrong," he says. He did more than prove his doubters wrong. He won many races and in 2008 was crowned champion in his age category.

With his love of automobiles, Mark thought he'd be a natural working for a car dealership. But there was the matter of his deafness. By the time he left his job as a social worker at Baycrest, he had very limited work experience, and most of it as a teen. Stints selling ice cream and delivering newspapers were successful; a brief turn collecting and storing shopping carts at a grocery store was not. He got fired because he couldn't hear the manager calling him on the public address system to pick up stray carts in the parking lot.

Mark was determined to succeed, but not by asking for accommodations or special treatment.

"I knew I had to be better than everybody else," Mark recalls. "I had to work harder. I had to be the first to arrive for work and the last one to leave. I had to give absolutely no reason for anybody to fire me. My productivity always had to be 10 percent better than anybody else's."

He managed to talk his way into a job as a service advisor at a major car dealership in Ontario. A service advisor is the person you first meet when you bring your car to the dealership for repairs. Mark knew his way around automobiles. What he discovered was that he could see things from the point of view of a wary customer.

"You've got to picture the way that customers look at you in the car business," says Mark. "They have to go to the car dealership. It's out of their way, they've missed their breakfast, and they know it's going to cost 500 bucks before the day is out. It doesn't matter how nice you are. You're the bad guy."

Mark developed an uncanny ability to get customers onside with repairs. But there was one big problem. Customers in the service department often don't wait around for estimates on repairs. They leave, which means it's up to the service advisor to call the customer up, make the case for costly repairs, and close the deal. For most service advisors, it's hard enough. For one as deaf as Mark, it's close to impossible.

"Today, I can't use the phone very well at all," says Mark. "Back then, I was able to use the phone to a certain degree, but it was dangerous because you're calling customers and asking if they'd like to go ahead with a $1,400 brake job."

With sketchy hearing, Mark couldn't tell for sure if the customer had said yes or no, so he began to ask his immediate supervisor to call the customer.

Mark's tenure as service advisor was financially successful. During the recession that struck the North American economy in the early 1990s, the service department where he worked was the only part of the dealership making money. But Mark had a nemesis: his immediate supervisor, the service manager. Eventually there was a showdown.

"One day, the service manager went to the owner and CEO [of the dealership] and says, 'I can't work with this guy anymore because he's deaf,'" Mark recalls. "He says to [the owner],

'I want you to make a decision: Get rid of him or I'm out of here.'"

Instead of firing Mark, the CEO gave Mark the manager's job.

Mark thought his boss's decision made economic sense. "Despite my significant disability, I was bringing something of value to his business. I couldn't answer the phones, but I had other people do that."

Back then, and maybe even today, that sort of accommodation was exceptional. Mark admits that closing the deal by phone was a key part of the service advisor's job description. Instead of accommodating Mark, the CEO could have agreed with the manager that not being able to close the deal on the phone for a $1,400 brake job meant he couldn't do the job.

Soon after the promotion, the phone problem became moot, as email became widely available. Still, he never forgot the CEO's willingness to accommodate him and the man's kindness.

"What I faced as a young man is the reason why I have empathy toward people with disabilities," says Mark. "All of society discriminates against them. It doesn't matter if you're deaf, blind, or in a wheelchair. I witnessed it first-hand, and I saw it happening to others. And that's why I knew I was going to hire Clint the moment he walked through that door."

Clint was the first employee Mark himself hired at Tim Hortons and the first with an obvious disability. That he was a smashing success encouraged Mark to hire another. And another. Today, 20 percent of Mark's employees have a disability. They work throughout his chain of restaurants. At Kennedy Commons shopping centre, his busiest location, he takes me into the kitchen to meet a young woman named Heather who has a cognitive disability and has been working for him for seven years.

"Ten years!" Heather quickly corrects her boss, with evident pride.

Mark flashes a smile. He likes employees who speak up.

"Heather went to school locally," he says. "She has a job coach who drops by periodically to see how things are going. But she doesn't need it."

"No, I don't," says Heather, shaking her head as she smiles.

"Heather comes in four days a week from 9 A.M. to 3 P.M.," says Mark. "She does the same job that Clint does, looking after the dining room and the dishwasher."

"I don't think I can remember the last time a supervisor came in to talk to you about your work," Mark says to Heather.

Besides Heather, the store has hired a number of students with disabilities who are enrolled in co-op programs. "One of the things we've discovered is that the number one indicator of getting a real job after you graduate from school is if you had a job before you graduated from school," Mark tells me.

Mark introduces me to James, one of two students with disabilities that Mark has hired through co-op placements. James is 24. He's doing a two-year Hospitality Management-Hotel and Resort diploma program at Centennial College in Scarborough. James has a cognitive disability, but he's functioned well on the job.

"I've been trained to work in the back of the store baking muffins and all the other food we bring to the front," says James. "I like the fast-paced environment."

When I ask Mark if employees with disabilities like James and Heather slow down the operations at the store, he gives me an incredulous look. "This store has just over 3,000 customers per day," he says. "It's our busiest store."

I can see where Mark's affinity for employees with disabilities comes from. But I want to know what his other employees think, so he takes me to a Tim Hortons located in an Esso service station at the corner of Lawrence Avenue East and Kennedy Road. When we arrive, the place is hopping with teenagers buying lunch. In the kitchen a woman is stacking boxes. Sazia Varachhiya is a 24-year-old woman who has been working for Mark since she was just 18. She is the store manager.

"What was your fastest time a couple weeks ago?" Mark ambushes Sazia.

"Fourteen-second window time over an hour," replies Sazia with a big grin.

"They're in first place on weekends," approves Mark. "During the weekdays, they're in first, second, or third place out of 500 stores in the Greater Toronto Area."

The location has a substantial complement of employees who require accommodations. "We have a number of people here who have intellectual disabilities," says Mark. "Employees with Down syndrome, employees with autism working in entry-level positions. We've got people working here in the drive-through on the weekends who have disabilities," says Mark. "She still gets 14-second drive-through times."

Sazia's most challenging employee is a man in his thirties named Lucas. Like Clint Sparling, Lucas has Down syndrome. Lucas cleans tables and empties the garbage for Sazia. She admits that it took her a while to adapt to Lucas.

"Honestly, it's sometimes really hard at first to understand people like Lucas properly," says Sazia. "Sometimes, it depends on their mood. I need to kind of set myself in their situation and try to sort things out—to go to their level, and then talk to them on their basis."

Sazia made a significant difference in Lucas's life. "Lucas spent much of his adult life living in his parents' basement, until he got a job here," says Mark. "Now, he's living in a house with three other men who have intellectual disabilities. He's a contributor to the economy. He's a taxpayer, and he's living a full life. This is a person whose support workers never believed would ever be able to work in a real job. Lucas proved them wrong."

The same is true of most if not all of the close to 130 people with disabilities that Mark has hired since he bought his first Tim Hortons more than 20 years ago. Each hire came with the challenge of figuring out what accommodations were necessary to make things work.

"One of the things that we started to do here years ago is what we call ATP," says Mark. "It stands for 'Ask The Person.'" Take, for instance, someone who applies to be a baker and just happens to be as profoundly deaf as Mark. "Your first thought is that it won't work because all the ovens have audible alarms," says Mark.

But the man who has made a success hiring people with all kinds of disabilities has learned to trust the person who applies for the job. "They're not going to waste their time applying for a job that they know they can't do," he says. "So, we ask them how they would get around their particular challenges. Nine times out of 10 they come back to you with something you haven't even thought about, and we end up hiring them."

Mark gives lots of speeches about hiring people with disabilities. At one event, a young woman named Risa approached Mark looking for a job. Following her graduation from university, Risa approached Mark at an event. Like Mark, Risa is deaf. "I want to come work for you at Tim Hortons," she told Mark.

Mark hired Risa to work in customer service. With orders from customers appearing on a computer screen, deafness was not a challenge. The job was only part-time, but Mark was impressed by Risa's attitude, in part because the young woman travelled five hours a day to work three hours a day at the donut shop.

Risa saw an internal posting for a full-time job as a baker and told Mark she wanted to apply. Mark told Risa that the ovens had audible alarms, but he didn't discourage her from applying. "I told her to take an hour, go into the kitchen, and figure out how to do the job," Mark recalls. Risa returned from her scouting mission after a few minutes.

"When the warning buzzer goes off, it's time to take the product out of the oven," Risa told Mark. "When the timer gets to zero, that also means it's time to take the product out of the oven. So what's the buzzer for? I get it! The buzzer must be for bakers who are too lazy to notice the timer has gone to zero."

Mark gave her the job. And the store's profits went up.

"Risa replaced a guy who had been working there for nine years as a full-time baker," says Mark. "I did an analysis of Risa's work. After a couple of weeks, her productivity was 18.4 percent better than the guy she replaced. By hiring somebody with a profound disability, I made more money."

Mark says experiences like that are the rule, not the exception. "I've hired 127 people in 20 years with a disability. Every one of them came through the door with a challenge, be it intellectual, physical, or emotional. In every case, the person turned out to be capable of more than I had assumed."

That lesson was reinforced not too long ago. Mark hired a man who had arrived from Ecuador two years earlier. The man had autism and other disabilities. "He wouldn't look you in the eye," says Mark. "He responded to everything with a yes. It didn't matter what you asked him; his answer was always yes. He had to be told many times how to do his job properly. I thought this guy was unemployable."

After two weeks, Mark was ready to fire the man. One of Mark's mangers wasn't so certain. "I've seen a difference in two weeks, just a small difference, but I have seen a difference," the manager told Mark, who agreed to continue the man's training.

"Fast-forward two years, and he was talking to all the customers, looking them in the eye, and holding conversations," says Mark. "He walked with a purpose. He no longer has behavioural issues. I had made a snap judgment call on his ability. I was wrong."

Mark says he was persuaded to be more patient with the man. But it doesn't mean he patronizes employees like Lucas, Risa, and Clint. "My expectation is that they work to the best of their ability," says Mark. "For some of them, their maximum ability may not be the same as everybody else's. That's okay. We make sure that the job that they're doing is the right fit for them. But they still have to bring something of value to my business. I'm not a charity. I'm not a babysitting service. I'm not a kindergarten."

It's a business, as Mark points out to me repeatedly. And a very profitable one at that.

"All six of my stores are outperforming the average among stores for transaction increases and sales increases over the last year," says Mark. "We're not better store operators than everybody else. We don't have any new factories that grew in the last 12 months with thousands of workers added to the payroll. We believe it's because we're an inclusive employer."

In health care, we're conditioned to think of patients as a drain on the system. With that mindset, it's not surprising that we see patients with a disability as an even bigger drain. The health care system hasn't gone as far as Mark has to make accommodations. It would never even occur to us to think that people with disabilities might be less of a drain on the system than people with no challenges.

Mark thinks we should consider his experience. "If you look at absenteeism, my 50 current employees with a disability have an absenteeism rate 85 percent lower than the 200 who don't."

Employee safety is another advantage Mark likes to crow about. His in-store safety rating is at the highest level with the Workplace Safety and Insurance Board, the agency that regulates safety and pays compensation to injured workers. "In 20 years with close to 130 people hired with a disability I've never ever had to fill out a WSIB claim for somebody with a disability," says Mark, who thinks he knows why.

"Other than a risk taker like me who drives racecars, people with disabilities don't take unnecessary risks," he says. "The person with cerebral palsy or the worker in a wheelchair is not going to climb out of that chair and get up on a stool to get something off the top shelf. They're going to ask somebody else to do that for them."

Employees with disabilities are safer, says Mark. He says their challenges also force them to find innovative ways to do everyday things. "If you use a wheelchair, you have to get up, take a shower, make breakfast, get in the elevator to go down to the car, and drive to work," says Mark. "Before you arrive for work, you've already demonstrated better problem-solving skills than you and I. That's how innovation is created in the workplace."

An astonishingly low turnover rate among Mark's employees is another huge factor in his success. "For a Tim Hortons or any quick-service restaurant that's well run, the typical employee turnover rate annually is 120 to 125 percent," Mark explains.

A 125 percent turnover means that for most other franchisees, the staff turns over completely in less than a year. At Mark's restaurants, the overall turnover rate is just 38 percent, a huge difference to the bottom line.

"Each new employee costs me about $4,000," says Mark. "If my turnover is 38 percent, and a fellow franchisee is doing just as good at the job but with a turnover of 100 percent, then I'm doing better."

Mark's overall turnover rate may be 38 percent, but among employees with a disability it's effectively zero. In other words, the 50 or so employees with a disability who work for Mark this year—people like Risa and Clint—are the same 50 who worked with him last year. Many of them plan on staying forever.

And the turnover rate among Mark's employees who don't have a disability is an enviable 55 percent, almost half the rate at other fast food restaurants. I can understand why people with a disability might want to keep working for Mark. But that doesn't explain why people who don't have a disability want to stay as well.

Sazia Varachhiya is one of them.

"I guess I'm a people person, and I'm very empathic toward people," says Sazia. "So that's probably why I like it here. What Mark is doing is obviously great for everyone. It's great for employees, and the ones who have disabilities."

She says her customers love Lucas, her employee with Down syndrome. "They love to see him. They talk about him, they ask for him, and they want to see more of him. He's very friendly and outgoing. When he sees customers, he goes out of his way to do anything to help them."

"We get emails and phone calls every day from customers asking are we the Tim Hortons that hires people with disabilities," says Mark. "We know there are people coming into our stores on a regular basis because of that."

Today, Mark is convinced that hiring people with disabilities is the way to corporate success, an idea he spends hundreds of hours a year promoting to business colleagues. In 2009, he persuaded Tim Hortons' parent company, the TDL Group Corporation, to launch a franchise-wide educational program on the merits of including people who have a disability. He also developed a branded decal that invites people whose disabilities make it difficult for them to order food at the drive-through station to proceed directly to the pick-up window for faster service. Mark persuaded the TDL Group to put the decal at every drive-through location across the chain.

Mark has met with finance ministers, provincial premiers, and even prime ministers. And he's gotten political action.

In the past, people like Clint and Lucas would have worked in sheltered workshops. The term refers to a workplace that employs people with disabilities in a location separate from people with no disability. Many sheltered workshops were set up following the Second World War to help returning soldiers receive rehabilitation and retraining to re-enter the workforce. As the number of veterans requiring such facilities declined, sheltered workshops were transformed into places for people with intellectual disabilities like Clint. "These places are completely secluded," says Mark. "The only people working in them are people with intellectual disabilities."

Paradoxically, people at sheltered workshops received lots of training. What they didn't get is a living wage, in part because provincial employment legislation exempted sheltered workers from receiving full pay. "Sometimes they'd get a stipend of a movie pass at the end of the week, but they never got paid," he says. "It's discrimination against people with disabilities. Without a proper wage, Clint Sparling would never have been able to get married and to purchase a condo."

In 2015, the Ontario government announced plans to close sheltered workshops forever. Mark is proud of the role he played in their demise. "I was instrumental in putting an end to the sheltered workshop model. I've been fighting it for 10 years, and I was largely successful the first week of December 2015 in getting rid of it."

He did so not only by speaking out against sheltered workshops but also by demonstrating through the success of his six Tim Hortons restaurants that people with intellectual disabilities can and should be gainfully employed.

And he's not done. Mark's next target is the one in five Canadians who has a disability, half of whom are unemployed.

"We believe at least half of those people could start work tomorrow," says Mark. "In Ontario alone, it's costing us $11 billion a year to support those people. You take just 5,000 of those people, put them into a combination of part-time and full-time work. You'd save the province $42 million per year every year because you take them off of benefits, and you create a new taxpayer."

The challenge is to convince fellow business owners that employing people with disabilities will make them money. Some, like the U.S. pharmacy chain Walgreens, get it. In 2002, Walgreens' senior vice president of supply chain and logistics, J. Randolph (Randy) Lewis, got the company to work with local agencies to train and attract people with disabilities for employment, including special training for managers. "Thanks to my good friend Randy Lewis, Walgreens is probably the most inclusive company in the world for people with disabilities," says Mark.

But Randy Lewis is the exception. Many CEOs have told him they won't hire people with disabilities. "You can't change somebody's attitude," says Mark. "You've got to wait for the next generation to come along. When they realize that by hiring people with disabilities, they can have a competitive advantage, they'll come around."

It's easy to say that Mark Wafer is empathic because it makes his donut shops profitable. It turns out that Mark's capacity to be kind to people with disabilities is off the charts. He demonstrates it by imagining the world from the point of view of a person with Down syndrome, like Clint Sparling, or a person with intellectual disabilities, like James, or someone with ASD, like the man from Ecuador.

Mark doesn't teach his donut shop managers to be kind. What he does is create a workplace that attracts kind managers like Sazia Varachhiya and Jennifer McCall.

Would he have those perspectives if he hadn't been born deaf? Probably not, says Mark. He doesn't think it was his family background that made him particularly empathetic. Mark chalks it up to the deafness. "It's the fact that I've had to deal with a disability that has given me a completely different outlook on life and on people."

It's not deaf people for whom Mark has empathy; it's deaf people whose capabilities are discounted. Mark felt the sting of rejection and used it to eclipse the doubters and to fight for the doubted.

And if you doubt that, Mark has one more story. It happened 19 years ago, when Clint was a new hire. The store where Clint worked had a periodic inspection by the city's health department. As the inspector completed the paperwork, Clint found out that the store had received a passing grade. "You wouldn't give us a pass if you saw the freezer on the other side," said Clint to the inspector.

Mark, who was there at the time, recalls that everyone including the inspector chuckled at Clint's innocent faux pas. The only one who didn't laugh was the store manager.

After the inspector left the donut shop, the manager confronted Mark. "I can't do this anymore," the manager told the donut shop boss. "You need to make a decision. It's him or me."

"Clint brings more value to my business than you do," replied Mark to the manager. "He comes to work early, and you don't. You don't take a bullet, and he takes too many. I can't get him to go home at the end of the shift, and you're in a hurry to get out of here. He works diligently all through the shift, and he's more productive than you are."

Clint stayed. The manager left.

"It was the right decision," says Mark, looking back. "It also showed everybody else in the store that I was serious about hiring people with disabilities."

Mark stood up for Clint, just as the CEO of the car dealership stood up for Mark many years ago. The donut shop owner had empathy for Clint. You might call what Mark did a form of payback. I prefer to see it as Mark demonstrating that people with disabilities want and deserve to work.

Something that looks cruel at first can turn out to be kind in retrospect.

In August 2017, after this chapter was written, Mark and his wife, Val, decided to sell their franchises. "The reason we sold was simply based on a sound business decision," says Mark in an email. "Our restaurants were particularly successful and that made them valuable to a buyer. After almost 25 years in business, it was also time to move on to other things."

His success as an inclusive employer at Tim Hortons has led to a second career advising policy makers and delivering keynote speeches to corporate, government, and service sector leaders. He continues to speak to audiences eager to hear all the reasons why hiring people with disabilities is good for business.

As part of his exit plan from Tim Hortons, Mark says he wanted to ensure that "all existing staff were guaranteed their tenure, rate of pay, and position with the new owners. This was successful, and all 46 workers with disabilities remain gainfully employed."

On January 1, 2018, the hourly minimum wage in Ontario went from $11.60 to $14 an hour, and in 2019, it will rise to $15 an hour. Ontario is not the only province doing this. Alberta plans on raising its minimum wage to $15 an hour by October 1, 2018, and British Columbia plans on increasing its minimum wage to $15 an hour by 2021. Early in 2018, the Bank of Canada said that these increases in the minimum wage could end up costing the Canadian economy nearly 60,000 jobs.

Ontario's minimum-wage increase led some Tim Hortons franchise owners in that province to reduce employee benefits and cut back workers' paid breaks. That led Ontario Premier Kathleen Wynne to conduct a media war with franchisees. In an election year, some have suggested she is trying to convince voters that she's on the side of workers. Others think she enjoys a public scrap.

Although Tim Hortons caught much of the adverse publicity, the minimum-wage increase affected other fast food franchises as well as other non–food service jobs.

A non-disclosure agreement prevents Mark from commenting on the donut franchises he left behind. But back in 2017, he says, he was in the thick of the controversy. "I encouraged the province to delay the implementation of the minimum wage," says Mark. "My hope is that all business owners will value the contribution their employees bring to their business and think hard about any decisions that will affect them. But the government of the day should have been far more strategic about launching this piece of legislation and allowed business to plan properly. I actually believe minimum-wage workers would have been better off overall with that strategy."

Sometimes, an act of empathy, such as trying to give workers a higher standard of living by raising the minimum wage, can have unintended consequences that look less kind in retrospect.

CHAPTER FIVE

The Bar at Ground Zero

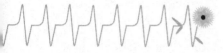

On an overcast morning in September, I stand in the middle of a crowd roughly 15 people wide and 25 deep. We stand silently smack in the middle of the intersection of Greenwich and Cedar Streets in Lower Manhattan—young families, middle-aged men in ball caps, tourists in tank tops. A white-haired man in a pale blue shirt and khaki pants, and wearing a small knapsack, locks the fingers of his hands and looks down. An Iraq War veteran with his head shaved quietly tells stories of his two tours of duty to his travel companions.

Ahead of the crowd is a 50-metre stretch of Greenwich Street bounded by barricades on each side of the cordoned-off road. At the end of the block, a phalanx of Fire Department of New York, New York Police Department, and Port Authority Police stand at attention with their backs to us. Just beyond the officers, a slow but steady procession of honoured guests makes its way across the road to the plaza.

At 8:46 A.M., bells softly ring, in memory of the exact moment American Airlines Flight 11 struck the North Tower of the World Trade Center. It's the anniversary of 9/11, the day when

19 al-Qaeda terrorists hijacked four U.S. passenger jets, two of which were crashed into the North and South Towers of the World Trade Center.

The mostly silent procession goes on for nearly four hours, until the names of all 2,983 men, women, and children killed in the attacks at the World Trade Center site are read and remembered by loved ones. As the day marches on, many of the invitees make their way slowly back to the corner of Greenwich and Cedar. They head to a bar like no other. It's a place with a wounded yet still beating heart that mirrors perfectly those of the patrons it serves. A place that empathizes with those who lost family on 9/11, especially the first responders who ran toward danger that day. A place that serves empathy along with the beer.

O'Hara's, a traditional Irish pub, sits right across the street from Fire Department of New York (FDNY) Ladder Company 10 and Engine Company 10, the only fire station inside Ground Zero.

The pub is just starting to fill. To my left, families in their Sunday best sit at high tables by the window. Just ahead, three bartenders wearing jeans and blue T-shirts serve drinks to a growing swell of active and retired FDNY firefighters, lieutenants, captains, battalion chiefs, and fire marshals in dress uniforms. Some have taken their jackets off, crisp white shirts and thin, dark ties underneath.

Immediately above the bar, an American flag drapes the ceiling. Green electric lights give the wood moulding that borders the ceiling above the bar an eerie glow. As I move closer, something else catches my eye. Embroidered insignias of firefighters, paramedics, police, and other first responders cover the wood moulding and almost every bit of wall space in the bar. Seven thousand patches and counting. A tall man of about 60 years of age notices that I'm looking at the patches. "That's mine over there," says Paul Fisher, pointing to a crest that bears the insignia of the Monmouth County Sheriff's Office in New Jersey. "I put it up two years after the buildings went down. Like all the other ones, it shows that we were here in this area at that time."

The day the Twin Towers fell, Fisher and his son were working in the county sheriff's office in New Jersey. They were two of thousands who were summoned to help clear the massive pile of rubble surrounding the place where the World Trade Center had stood.

"That was a Tuesday, and we were down here by Friday morning when President Bush was down here at Ground Zero," Fisher recalls. "We were the Bucket Brigade. We were loading the buckets up and handing them down throughout the line. We were on top of the Pile."

The Pile is the term used by rescue workers to describe the tons of wreckage that remained from the collapse of the World Trade Center. *Bucket Brigade* refers to the thousands of volunteers like Fisher who passed 20-litre buckets full of debris down the line to investigators. Just 20 survivors were pulled from the rubble, the last one 27 hours after the buildings collapsed. After that, the objective was to recover bodies.

"Any bucket that contained a body part was labelled with a ribbon that they had to tie on," says Fisher. "As the bucket went by, you knew a body part was in there. At the bottom of the Pile, they would separate body parts from the other debris."

Fisher and his son worked for three days without sleep. "Adrenaline just kept us going," he says. "We were here all day Friday and went home Sunday night."

Fisher returned to O'Hara's on the first anniversary of 9/11; he's been coming each anniversary ever since. Fisher got up this morning at 5:30 to get here early. He and hundreds like him. "From all over," says Fisher. "I don't know how they found out, but I guess it's because we all gather down here in Lower Manhattan, next to Ground Zero. This is the closest bar to Ground Zero in the area."

Like Paul Fisher, Robert McGuire returns to O'Hara's each anniversary. But McGuire's connections are a lot closer to Ground Zero. The retired firefighter was on his way to work at Ladder 10, Engine 10, when the first plane struck.

"I got here as the second tower collapsed," says McGuire. "We were over by Trinity Church, which is one of the few spots around that didn't get crushed. And we were lucky enough to be there when the North Tower fell."

Being there—steps from Ground Zero—is an experience that has stayed with McGuire.

"It was eerily, eerily quiet," he says. "All you heard were alarms going off. No sounds other than that. It was the quietest I've ever heard New York City. It looked like it was snowing. Paper was still falling and dust. The ground was covered with a couple inches of dust."

He and the firefighters with him moved into what remained of the World Trade Center.

"We started climbing over what was left of the Twin Towers, which at some points was 10 or 20 storeys high," he recalls. "It was basically just twisted metal and debris. You didn't see one door, one toilet bowl or one desk. You saw nothing that was in those buildings. Everything was just turned to dust—steel and dust."

McGuire knew that more than 300 of his colleagues had already responded to the initial emergency calls and were inside when the towers collapsed. "You had no idea where any of them were," he says. "In the early days, we were all hoping. There were certain areas that weren't that affected in the courtyard, and we thought maybe there might be places where we'd find people later on. That was the hope for the first few days until it became quite apparent there would be no survivors. I lost over 75 friends and colleagues that I worked with that day."

That total includes McGuire's nephew Richard Allen, a 31-year-old probationary firefighter at Ladder 15. "He was six weeks on the job," says McGuire, looking away from me. "It was his very first fire."

McGuire was on the job clearing rubble the day his nephew's remains were found. "He had a tattoo of two dolphins swimming in a circle," he tells me. "And that's how we identified him."

McGuire retired later that year after knee replacement surgery. He doesn't talk about his feelings. Coming back each year on the anniversary of 9/11 is his therapy. "It's to pay tribute to all our lost brothers and all the people killed that day," he says. "We go to the memorial and touch the names of all the people that we know, which is many, many people."

After that, he comes to O'Hara's—every year. "The people here are great," he says.

I ask the retired firefighter what kinds of conversations he's had with the bartenders.

"None yet," he says. "We stay pretty closed ranks with our brother firefighters."

So what makes this place special for McGuire? Like Fisher, he points to the insignias that fill the walls. "As you can see, row after row of patches of all the various companies from around the world who come to pay tribute," McGuire says.

People like Shawn, a 26-year-old paramedic from Gary, Indiana. "I can pick up a conversation with anybody here and I'll talk to him for two hours," he says. "We only came here for one drink the other day, and we stayed for five hours. I met a firefighter from Scotland. I don't know why the hell he came from Scotland, but God bless him."

First responders are not the only ones who flock to the place, says Paul Fisher. "I've seen airline pilots come here because they were flying aircraft on that day. Flight attendants too. You will find a lot of family members. They come in with photos of the loved ones that they lost. This is the spot where everyone comes. In another two hours, you're going to find out that you can't even get a seat in this place."

Fisher is right. The large tables by the wall are filling up with families who took part in the memorial service.

I notice a man in his fifties as he walks slowly from table to table, talking with the families. Between tables, he greets the horde of first responders with strong grips and hugs. Paul Mackin turns to face me as I walk toward him. He looks weary— just like the first responders and health professionals who work all hours. Just like me. Sad eyes and an open face that suddenly breaks into a wide grin.

"Welcome to O'Hara's," says Mackin, the co-owner.

I'm here to meet Mackin because I'm told he's the kindest bartender at the kindest bar in the world. Mackin sits me down, offers me a pint and a story.

"A guy I'd never met before was here with his three daughters," says Mackin. "He lost his twin brother—a firefighter—in 9/11. He was here with his wife and he hung his brother's patch up in the bar. He was so proud. I met his daughters. We started talking, and his daughters pulled me aside and said, 'You know you made his day. He's so happy.' That's what we do here. We offer hope for good people."

Mackin has a near photographic memory. Every inch of every wall in this place has an artifact with a story. He points to a flag and a name tag by the bar.

"That's Adam Keys's name tag up there," he says. "He was in here on the Fourth of July and we sat down and we talked. He looks at me and says, 'I guess I've got to give you my patch when I retire.' Then he pulls up his pants, and he's got no feet, no ankles, and no arm. I'd have never known. He had his legs blown off and his arm blown off in a roadside bomb fighting for this country. He was the first triple amputee to go back to full active military duty. And the day he retired, he drove three hours from Pennsylvania to hand me that flag and sign it, and I hung it up in my bar proudly."

For some, Mackin offers hope and a bit of validation; for others, it's impromptu counselling. Mackin tells me about a firefighter named Sean who was a "chauffeur "— slang for the driver of a fire engine.

"It was about 9:30 in the morning on one of these 9/11 anniversaries," Mackin recalls. "Sean was in the bathroom crying. I asked him what was wrong, and he said, 'It's too difficult for me. I drove these people to their death. I was the chauffeur.'"

What the heck does a bartender say about that?

"'Yes, Sean, you were the chauffeur,'" Mackin recalls telling Sean. "'You did your job. You stayed with the rig. They did their job and tried to save people. What you need to do is live your life because you're still here. God spared your life for some reason, so you better figure out what it is and you better go live it because the guys who went the other way would want you to go live.'"

Did his pep talk work?

"He was okay," says Mackin. "He heard me. It's still difficult for those guys but that's what you do. You try to help people through their lives. That's what this business is about. Come in as a stranger, leave as a friend. That's what we do here, and we do it well."

Mackin certainly does it well. I watch him work behind the bar for a while. I see patrons in twos and threes talking among themselves. At opposite ends of the bar, two young men sit quietly, talking to no one. Mackin has a brief chat with a group of three seated next to the first guy. In the blink of an eye, the three move over, making room for the young man at one end to sit beside the other. They don't talk to one another. That's not the point.

"A good bartender manages his customers from the other side," says Mackin. "He orchestrates. He drives the bus, and he has to make sure the right people sit next to each other. He has to make sure the two quiet guys with their computers sit next to each other and neither wants to talk, but the crowd who wants to talk, you sit them next to each other. Then the job does itself."

Mackin can keep track of the moods of 20 customers at a time. "You orchestrate the environment," he says. "If the environment is positive, then the job is positive. If the environment is negative, you've got to figure out why it's negative. A gifted bartender curtails that negativity and turns it around."

Mackin tells me how he begins to do that. "You find out where they're from," he says. "You find out how their day was. You find out what they're doing here. Why are they sitting in front of me? That is the question I ask most people. I can usually tell if something's wrong by looking at them. Sometimes they tell me."

And if they don't want to talk?

"You leave them be."

Until they've had a couple of drinks.

"Alcohol is basically a truth serum," he says. "After several drinks, they reveal who they really are. You'll have an angry person. You'll have a sad person. You'll have a happy person. You'll have a violent person. It's all in stages."

The bartender says he has helped customers through the worst of crises.

"I've got a customer who went through a terrible divorce," Mackin recalls. "I was here with him as he was reading text messages that were coming from his wife. He'd programmed her phone so he would receive all her text messages. She was texting her lover while he was sitting in front of me."

With each text and each drink, the customer became more and more enraged. Mackin cut the customer off and managed to calm him down, but he said nothing about the ethics of programming his estranged wife's phone. He didn't judge the customer, just like I don't judge my patients.

"A lot of times, a person who's drinking won't make the right choice," he says. "You have to make sure they make the right choice. Sometimes the right choice is not to have another drink and go home. Sometimes the choice is to have a drink and I'll make sure they get home. That's part of my responsibility, as a bartender and as a human being."

This bartender goes far beyond his job description.

"A guy worked as a roadie for a famous band," Mackin recalls. "He was dating a good friend of mine from Jersey. They were living in an apartment around the corner from here. I became friendly with them over the course of time, and their relationship broke up over his booze. He was a lighting tech and he had a good job. He worked for bands all over the country, travelled, always on the road."

But his drinking started to get out of hand.

"I'd let him pass out in the car," says Mackin. "My wife and I would drop him off on the way home."

Mackin says the breakup and other things made him re-examine his life. "He went to rehab, and he's been clean for two years," says Mackin. "He got his life back."

And that is when something wonderful happened.

"He came in to see me and we spoke," says Mackin. "Then she came in to see me and we spoke. She started crying when we started talking about him because she loved him. They're back together."

"Do you see yourself as a marriage counsellor?" I ask.

"Marriage counsellor, brother, cousin," he says. "You see a woman sitting here by herself and she's having a drink and she's here from out of town. You go out of your way to make her feel welcome, and to protect her so she can come in here by herself and not be treated badly by the other customers."

If that happens, Mackin steps in pre-emptively.

He sounds like a therapist. "I am. I just serve alcohol while I do it."

I ask Mackin if he was empathic as a child.

"I always had compassion for people in every stage of my life," he says. "I've held the door for a stranger. I'll pick a guy up off the street. I'm a human being and I was raised that way."

He says he learned about being kind from his parents, now both in their mid-eighties. "They are the most generous, loving parents you could ever have in your life," he says. "They taught us respect. The generations were different 40 or 50 years ago. You did what your parents told you to do. You didn't question it. You didn't have an iPhone. You didn't have all this technology that challenges the youth of today. I think the generation growing up now has lost some of their people skills because they're involved in their phone and their electronic lives."

As difficult as it is to imagine now, until the day before 9/11, O'Hara's Pub & Restaurant was not a therapy bar. It was just another traditional Irish pub in Manhattan's Financial District, or FiDi. The area, which occupies the southern tip of Manhattan, is home to the New York Stock Exchange, the NASDAQ, the Federal Reserve Bank of New York, and the headquarters of many banks and other financial institutions.

Mike Keane, co-owner of O'Hara's along with Paul Mackin, knows the bar's history. The thin but well-muscled man in his fifties greets me with a strong handshake. Paul Mackin lives to tell a story. Keane acts more like the bar *is* the story. He is a beehive of activity, slinging cases of beer between sips of conversation with me.

O'Hara's opened its doors in 1983. Keane joined soon after.

"Two guys opened the place in '83," says Keane. "About eight or nine months after they opened it, they asked me to work for them. It was after high school. I didn't really know what I wanted to do. I didn't feel like wasting time going to college if I didn't know what I was going to do. I was supposed to work on Wall Street because my father worked there, but I didn't care for that."

In 1985, Keane liked the bar business enough that he became a co-owner. "O'Hara's was what I liked in a bar. Lots of down-to-earth people," says Keane. "We'd have construction workers in, and we'd have bankers, guys in suits and financial industry people. There were no tourists down here at the time."

The most colourful customers were commodities traders. At the time, the Commodity Exchange, a commodity futures exchange that is part of the New York Mercantile Exchange (NYMEX), had its headquarters at the World Trade Center.

"Did you ever see the movie *The Wolf of Wall Street*?" Keane asks. *The Wolf of Wall Street* is a 2013 film directed by Martin Scorsese and based on a memoir by Jordan Belfort, a high-flying stockbroker who in 1999 pleaded guilty to fraud and related crimes in connection with stock price manipulation.

"That was these guys," recalls Keane. "Maybe it wasn't $100 million, but these guys were all big-time partiers. They'd made a load of money. It came in and it went out. These guys would drink like it was New Year's Eve. Every morning at 11 o'clock, they'd be in here."

Keane says a typical day rolled out like a parade. "The commodity guys would come in," he recalls. "They were loud, and they drank up a storm. The construction guys came in, and they fit in with them. They would start drinking early, and they'd leave at 4 or 5 P.M. Then, you'd get the more reserved Wall Street people coming in after that. It was like a good changing of the guard every day."

Keane remembers the regulars. "There was a small guy that would come down and say, 'Mike, give me seven dollars' worth of vodka.' Twenty years ago, that bought a nice glassful. One of the other guys sitting there would say, 'Mike, give him another seven dollars' worth.' He'd have two glasses of vodka and go back up to the floor to trade. Then the other guys would come down yelling at me, because he's up there stewed. I said, 'Hey, listen, it's not my fault.'"

Paul Mackin started working at O'Hara's in 1988.

"My father owned the building back then with Mike," says Mackin. "That's how we became friends and [eventually] partners. I came to do the vacations for everybody else who worked here. I had never bartended before. I went home after my first day of working here and I said to my father, 'I know what I'm going to do the rest of my life. I found my passion.'"

By passion, Mackin meant the people, the job, the service, and the money he put in his pocket at the end of the day. 9/11 was years away. "It was different because people were more carefree," Mackin recalls. "It was about living life to its fullest. I had guys I'd serve at nighttime that would show up and knock on the window at 9 the next morning—wearing the same clothes—and looking for a drink before they went to work."

Mackin worked at O'Hara's for 10 years. In 1998, he left to work for his father, who had opened up a restaurant in New Jersey. Though he kept in touch with Keane, it would be several years before Mackin returned to O'Hara's as a partner.

Then came 9/11. Mike Keane was already at work when the first plane hit.

"We had come early in the morning to get ready for the day," says Keane. "I was on the phone with one of the commodity traders at 4 World Trade, and he just said, 'Hey, Mikey, the lights just dimmed over here.'"

People started to pour into the bar. Keane went outside to have a look. "There was glass sort of littered up the street," Keane recalls. "We're south of the Trade Center. The North Tower had gotten hit, but from our perspective, you couldn't see anything. We're only a block away, but all the glass was there. They evacuated some of the buildings over there, so the commodity traders all came here. They thought they'd be out of work for two hours; nobody really knew what was going on."

Keane turned on a 19-inch TV set—the only one at O'Hara's back then. "You could see the hole in the tower," he recalls. "We were getting busy. All these guys were coming in drinking. The phone was ringing, and things were going on." Keane and a couple of co-workers went up to the roof to try to see what was going on. "It's a five-storey building, and some papers up there were smouldering. So, we put the fire out. On the way back down, the other plane came. You knew something was wrong. So, we got everybody out of here."

Keane and his co-workers went back up to the roof just as South Tower started to come down. "Who the hell thought the tower was going to fall straight down?" he asks. "We ran down the stairwell to the basement and hosed ourselves off. We were all covered in the dust and debris. As I was leaving, a cop reaches me and says, 'Help me look for survivors.'"

Keane went with the NYPD police officer searching for people to rescue.

"We were up the street in Zuccotti Park, looking under all the hot dog wagons, and everything was pitch black," he says. "The cop had a flashlight. We were looking under there when the North Tower came down. We took off and went into some building. I was in there for an hour until everything cleared."

He made his way on foot to his father's office on Broadway; the two managed to reach the river and got on a tugboat heading for New Jersey.

New York mayor Rudy Giuliani issued an order barring vehicles and pedestrians from entering Lower Manhattan, but Keane managed to return to O'Hara's the very next day. He took stock of the damage. "It was just dust and debris everywhere. The roof was damaged. Everything on the roof was damaged. So were some of the floors where the larger windows are, on the corner of the building."

Still, the basic structure of the building was intact. It would be six months before Keane could get the bar repaired and reopened. But that doesn't mean the bar served no purpose in the immediate aftermath of 9/11.

Robert McGuire, the firefighter who walked into Ground Zero just after the towers fell to look for survivors, remembers clearly the first time he walked into O'Hara's: "I think we might have 'liberated' it the night of 9/11. The place was intact. Of course, everything was candlelight. We didn't want to see those beers go warm so we had to drink a few of them. I think we might have 'liberated' a few cocktails out of the bar that night as well."

In those chaotic first days after 9/11, rescue workers commandeered O'Hara's as a gathering place. "It was pretty much our headquarters for a few days," says McGuire. "You had firefighters here. You had nurses here—a lot of them. Everyone came here to wash his or her eyes out because we were blinded with the dust. It was all lit by candlelight. You had people giving massages. It was like a triage place."

Mike Keane was in and out of the building taking stock of the damage. O'Hara's was in no shape to reopen as a bar. Still, it continued to be an informal meeting place long after first responders were given temporary digs. "Guys would come in on breaks," he recalls. "They would come in to see how we were doing. They would hang out for a while, and then go back to the Pile and back to work."

Paul Mackin got his first look at O'Hara's a week after 9/11. "Words could not describe what I saw," says Mackin. "They were afraid that if it rained and the roof got wet, it would collapse."

Mike Keane looked at the massive job of trying to reopen O'Hara's and considered closing it for good. "A lot of people asked me why we decided to reopen," Keane recalls. "We actually had a fire here in '92. Basically, we had to gut the bar from all the water damage and everything else. After that, you weren't as devastated with the destruction [of 9/11]."

He started thinking about people worse off than he. "There were people that were lost, or had lost everything. It was just a pretty miserable time."

Meanwhile, Paul Mackin was going through a mid-life crisis of his own. His dad got sick and sold the restaurant in New Jersey. Mackin was at loose ends. Coming back to help clean up the damage to O'Hara's after 9/11 cemented his friendship with Mike Keane.

"Mike kept calling me to ask if I'd come back as a partner," says Mackin. "So we sat down. We talked. I said, 'It's not like it used to be but, Mike, you survived a fire. In 1992, this place burned down. You survived 9/11, and you survived a bad economy. The only place you can go from here is up.' I was so overwhelmed that he thought of me to come back as a partner and ask me [to share] that opportunity." He came back as a partner in 2004.

Despite considerable challenges, O'Hara's reopened six months after 9/11. At first, there was scant business. Mackin says two things saved the bar. One was that the area surrounding O'Hara's became a residential neighbourhood, bringing customers seven days a week. The other was the decision by Keane and Mackin to make O'Hara's a special place for 9/11 survivors and their families. That began spontaneously on the first anniversary of 9/11, when a customer named Big John ripped a patch off a first responder's uniform and stapled it to a wall. Other customers followed suit. By the end of that first anniversary evening, there were more than 250 patches stapled to the wall.

Keane says the patches have had a snowball effect on visiting clientele.

"They're happy to come in, and they are honoured to put

their patch up next to all the other guys that have been here," he says. "They feel comfortable and wanted. All of the guys that come down to pay their respects to the 9/11 memorial find us. If you ask any of the cops or firemen where to go, they send you over here. A lot of the same guys from all over the country come back every year and meet up with other guys that they just met on 9/11, and they've become friends."

That includes retired firefighter Robert McGuire. He neither works nor lives anywhere near Ground Zero. Yet he comes back every year on the anniversary. "The fire department has its own memorial up in Riverside Park, but a bunch of us just feel this is the place to be. This is where it really happened," McGuire says.

The Brazen Head is a trendy bar on Atlantic Avenue in Cobble Hill, a 40-block neighbourhood next to Brooklyn Heights. It's a 20-minute cab ride from O'Hara's. From the large gaudy sign in brass-coloured letters on a blue background, it doesn't look or feel anything like the Irish pub where I've spent a good part of the day.

After my time at O'Hara's, my search for other empathic bartenders has brought me to this place. The walls are exposed red brick. Long-stemmed pendant lamps with orange-gold shades shaped like upside-down flower vases drop down from a high ceiling. Along the wall behind the long bar are six window-sized chalkboards crammed with food and drink specials.

The only two obvious things this place has in common with O'Hara's are beer (15 craft beers on tap) and a dart board.

The place is filled with 20-somethings. There are no uniformed first responders or sombre families. There's no 9/11 anniversary vibe whatsoever.

Katharine Heller is sitting by the bar. The slim, fortyish woman with medium-length brown hair is wearing a pink and black shift dress. Heller has been a bartender since she was 22. She teaches mixology at a school and hosts a podcast called *Tell the Bartender* on which everyday people tell her unique stories.

I started listening to her podcast and immediately got that sense that Heller is both empathic and good at noticing empathy in others. I figured I had to meet her.

Heller doesn't work at the Brazen Head; she's brought me here to watch and learn. I ask her to describe the mood in the bar.

"It's a very jovial feeling," says Heller. Her eyes do a systematic scan of the room as she speaks. "We have here people who maybe weren't even in the city during 9/11. This is a bar full of people looking to have fun."

I point out that it's 9/11 as we speak and we're in New York City. Heller mulls that over.

"There's a lot of people who aren't talking about it, don't want to be involved in that kind of conversation, and don't want to watch the news," she says. She trains her sights on the bartender, a neatly groomed man in his early to mid-thirties who moves quickly and almost robotically through his routine.

"I don't want to talk about it too much because he's standing right there," she whispers as she leans toward me. "He's nice. He's just doing his job on a regular shift. This is Sunday, which is a great day to work because people like to drink early and go home early because they have to work the next day. Sunday is my favourite shift because it's just a good group of people who come here to have fun, forget their problems, and literally just be like children."

"But he's not you," I reply.

"No, he's not me," she says. "I could be totally wrong but I find that men have not been taught that fostering a feeling of empathy is a positive thing. You're not 'manly' if you are a feeling person. I don't fault men for that. I think it's society that teaches young men that they shouldn't be in touch with their feelings. Women are encouraged to talk about feelings and men are not."

I ask her what empathic qualities she brings to the job of bartending.

"A good empath can sense basic energy and know [it] without even saying a word," she says. "The second a customer walks in, I always know what they are going to be like. I always know when to be worried for someone and when someone is not going to be a problem."

Heller pays attention to nonverbal cues. "You can tell by the way someone is standing, by the way they approach you, eye contact, and by the way they handle themselves around other people," she says. "It's kind of like when dogs are in a playground, and one of them is clearly the alpha. They don't use words. They just know by body language who is in charge, and you can sort of tell what someone is going through when they walk in the room."

Heller reads the gamut of emotions. Take sadness, for example.

"If you take a second to look in someone's eyes, you can see sadness even if they're smiling," she says. "Most people are too scared to look into someone's eyes for a long period of time. If they want to talk about it, I just let them know that I'm around. I might want to keep an eye on them to make sure they're not drinking too much, which is also a thing you have to worry about. I kind of feel responsible for everyone's emotions and everyone's experience at the bar."

That means knowing the difference between a customer who wants to talk and one who wishes to be left alone.

"A comedian from Comedy Central came in, and everybody at the end of the bar recognized him," she says. "I said to them, 'Guys, you are not to walk up to him. He clearly just wants to have a beer.'"

Heller is good at detecting the angry customer who is about to explode. "I know when I sit down next to someone that he is going to be a problem. If they come in angry at something, I'm hoping that they'll calm down after a drink. I don't want to be on the receiving end of why they are so mad. I have had fights break out."

She's become an expert at defusing tension.

"If a fight breaks out between two dudes, the female bartender breaks it up," she says. "If you send a male bartender out there, something [bad] is more likely to happen. A woman takes them by surprise. The first time, I was really nervous to do it. I came out and said, 'Hey, guys, if you want to take it outside, go outside. Or, I'll buy you a drink here and we can figure this out.' By the end of the night, they were hugging each other."

Like Paul Mackin, the bartender and part owner at O'Hara's, Heller has her share of stories of the times when she used her empathy to make a difference.

"Some customers of mine found out that their apartment was burning down while they were in the bar," she says. "I called them that night. I organized a fundraiser for them. We gave them toothbrushes and underwear and helped them clear things out. We got the whole neighbourhood together."

She doesn't do that for everyone.

"I'm careful to whom I give my personal information," she says. "If they're regular customers in the neighbourhood and I get to know them, their spouses, and their families, I feel like we become friends."

Nor does she dole out kindness to every customer. "There was a guy who was just being an ass," she recalls. "We realized he was a little too drunk. He was sexist and racist and calling us names. My colleague Lisa took money out of our tip jar and handed back the amount of the drink that he paid for. Then, she took the drink out of his hand and said, 'Here's your money back from our tip jar. This is insurance that you'll never come back.' Everybody at the bar applauded."

Heller says there's another way to know if the bartender thinks you're a piece of work.

"Cutting people off at 10 o'clock at night," she says. "You hand them water and then you just ignore them or give them terrible service. Invariably, they say that the bar sucks, and then they walk away."

Those protective manoeuvres stand in contrast to Heller's overall kindness. On the scale of empathic bartenders, she's off the charts.

"I can read energy," she says. "I know it sounds weird, but I can literally see your aura. I know within a second of someone walking into the bar if they're going to be weird or if we're going to get along perfectly. Sometimes, I don't want to believe it, but I'm always right."

As a journalist used to asking questions, I'm a bit jarred when the kind bartender turns her attention to me. She insisted on getting to know me by phone before agreeing to meet in person in New York.

"I needed to hear your voice because I wanted to get a feel for you," she says. "The second I got you on the phone I knew it was going to be fine. I love that you're learning about empathy and I love that you're doing this book. I guess anyone out there who is learning about this has the power within him."

It's far too early in my search for empathy to know if she's right. Then she says something that makes me feel as if she can see inside me.

"I don't want to sound rude, but empathy is something that is frightening to tap into," she says. "It means you have to take a good hard look at yourself and be aware of other people's pain and not just your own."

Heller knows what she's talking about. She's known that she's different since she was a little child growing up in the Bronx. "I thought something was wrong with me because I could feel everyone's energy all the time," she says. "As a kid growing up, I could sense when my mom was upset, sad, or anxious, and it made me anxious."

Anxiety was a constant companion when she was growing up. "I would get anxious because I was just feeling it," she says. "I have never known a single time where I wasn't a pretty anxious kid."

Heller says she had good reason to be anxious. Her parents divorced when she was a child. It was her father's second marriage. She has two half-siblings from her father's first marriage. She says he was in and out of her life. "I've spent a lot of time not really trusting people, and I still don't," she says. "It's one of the things I work on in therapy."

Heller was diagnosed with obsessive compulsive disorder (OCD) as a kid. Medications and therapy helped her get a better handle on her OCD and her anxiety. Gradually, she began to see her ability to sense the feelings of others as a gift.

"People tell me things they shouldn't," she says. "I feel for someone when they're going through pain. I've had to learn to build a wall, so I don't get affected when a customer comes in with really sad news or something like that, because there's only so much I can take."

After finishing high school, she studied communications and theatre at a liberal arts college in Minnesota. After moving back to New York, she needed to pay off her student loans. She applied for a job as a cocktail waitress at a neighbourhood bar. She was asked if she'd ever been a bartender. She lied and said yes, and so they hired her. That day, she bought a book and taught herself the basics.

"I found that it was a lot more lucrative than an office job—with a lot more freedom," she says.

Heller got proficient at it and began to teach others. She says she has empathy for newbies. "I do a crash course for people," she says. "If it's just a regular bar, you only need to know about 25 drinks. I teach people how to look like they've bartended before so that when they lie on their resumé—and they all do—at least they don't look stupid behind the bar."

"Are bartenders supposed to be empathic?" I ask.

"A million percent," she says. "There's this idea that bartenders are supposed to be gruff, sassy, and rude. Truthfully, those are not the ones that get tipped well or have returning customers."

I ask if she thinks she can teach empathy to bartenders.

"Yes and no," she says. "I think that most people have empathy. I also think there are some people who are just hopeless."

Heller doesn't just tend bar. She also frequents bars. I wanted her take on what makes a good bartender.

"I don't care how good the drink is," she says. "If you're an asshole for no reason, I do not want to go to your bar. You don't have to be the nicest person in the world. I don't need to be coddled and told I'm special. Literally, I just need you to say, 'Hi, what would you like?' That's about it. If you're just rude to me, I'm not going to like you."

She has an interesting way of telling rude bartenders how they did. "I always just tip really well," she says. "I make sure to look them in the eye, and I walk out. I never say you fucked up, I just say, 'Here's your tip.' And they know. It's a hint. It's money that burns because no bartender wants to earn it that way. It's shameful. We can be really cruel. It's a weird power play. I hate doing it but part of me thinks they should know."

At the back of the bar next to the dart board one of the bartenders grabs a microphone.

"It's almost that time," says the bartender at the microphone. "That's right: a very special Sunday edition of Trivia Challenge."

"I'm sorry about this," says Heller. "When I heard the loudspeaker, I thought for a moment they were going to say something about 9/11."

Me too. A few kilometres away, they're commemorating 9/11 at O'Hara's. At the Brazen Head, they're doing a trivia contest.

Suddenly, 9/11 is all Heller wants to talk about.

"People who were there during the thing are still in shock," she says. "On the anniversary, we don't talk about it with each other. I never talk to people about 9/11 *ever*. I never email them to ask how they're doing. Usually, I am home by myself. I still can't take the subway on that day."

I ask what it was like to be a New Yorker back then.

"The days after 9/11, when the subways started running again, people on the train were so spooked," she says. "One time, there was a woman I was staring at because she looked like she was in shock. Everyone had PTSD, and then they would have to get on the train and go to work. This woman was staring at a subway poster of the World Trade Center. She was dressed in a business suit but she looked like she was about to crumble. That's when I realized everyone in this city is scared. There was another time when somebody ran out of the subway car because they had gotten off at the wrong stop. Everyone thought it was a bomb threat, so everyone ran off the subway."

She scans the room at the Brazen Head and comes to a conclusion.

"There are a lot of people here who were not in New York during 9/11," she says. "This neighbourhood was in the direct line of smoke and ashes. Debris from the World Trade Center fell in my friend's backyard right down the street from here. The fire was burning for four months. We smelled bodies burning for months."

Exposure to the smells and sights and other detritus of 9/11 is only part of what New Yorkers are trying to block on this anniversary day.

Paul Mackin feels it. Sean, the firefighter who chauffeured his comrades to their deaths, feels it. So too does retired firefighter Robert McGuire, who lost 343 of his fellow first responders including his nephew.

It's called survivor guilt, says Heller.

"I know an FDNY lieutenant whose name is Tom," she says. "He is an amazing person who lost all of his men. He went down to Ground Zero [during 9/11]. He was really measured, but I know he felt terrible about that."

She remembers Tom telling her that he was diagnosed with PTSD and forced into retirement. "He got a huge settlement. He got millions of dollars because he worked down there. He just took this money and left New York City. He said to me very plainly, 'I can't work anymore. I'm very sick in the head. I'm emotionally damaged.'"

First responders are not the only ones to be wracked with survivor guilt, as she found out in the days immediately following 9/11.

"There was a guy who used to come to my bar," she says. "He was a banker who always dressed in a suit. He was the nicest person in the world."

Heller says the man was working at the World Trade Center on 9/11 and managed to escape after the plane hit the tower he was in.

"He started coming to the bar every night drinking himself into a stupor," she says. "He never had a problem with alcohol before. I gave him my phone number and I said to him, 'My mother is a psychologist. I can help you find someone to talk to. I just can't serve you alcohol anymore because you're killing yourself.'"

Heller's intervention helped the customer enter recovery. "He ended up writing my boss a note in which he said that I saved his life," she says.

Suddenly, Heller remembers one more story. She and her sister had been friends with a firefighter named Christian. The young man was one of the 343 who died on 9/11.

"I didn't realize he was a firefighter from my neighbourhood," says Heller. "My sister doesn't like to say it, but she named my nephew Christian after him. She and I don't even talk about it. Yesterday, there was a block party, and there was a fire truck parked on the street. The truck had Christian's name on it. All the kids were taking photos of the truck, and their parents wanted to take a photo of my nephew beside the truck bearing his name. I went up to the driver of the fire truck and said, 'Hey, I actually knew Christian. Did you know him?' He said, 'No, I didn't.' We just looked at each other and we shook hands, and this grown man started tearing up."

Heller turns wistful. "Now, it's younger blood and new people," she says. "I'm sorry I just keep talking about it. This is the most I've talked about 9/11 ever."

9/11 may be fading into the background, but it's never far

from the hearts and minds of people like Heller. And it may not be as far away from newcomers as she thinks. A while following my visit, on October 31, 2017, a Home Depot rental truck driven by Sayfullo Saipov, a 29-year-old immigrant from Uzbekistan, sped for more than a kilometre down a popular bike path in lower Manhattan, killing eight people. Saipov was charged with providing material support to ISIS and with violence and destruction of a motor vehicle. The attack took place in the shadow of the rebuilt World Trade Center.

Is it kinder to not talk about 9/11 and just keep going? Or, is it better to give the day a voice and a place and to remain wary of the next such attack? To answer the question, I bid goodbye to Heller and head back to O'Hara's.

At 7 p.m., the sun is a giant ball reflected in the Hudson River as it sets over New Jersey. Thin wisps of orange light penetrate the buildings around O'Hara's. The 9/11 crowd has thinned considerably. Most of the families have left.

Some bartenders like Katharine Heller are born to be empathic. Others, like Mike Keane, have empathy thrust upon them by circumstance. I came back to ask Keane how 9/11 and the anniversary have changed him.

"It's tough," he says. "Starting yesterday, you get that sick feeling in your stomach. You think about seeing everybody today. There was a friend of mine that passed. His wife, Nancy, came in with her daughter. Nancy kept saying, 'Mike knew Dad.' And her daughter just kept saying, 'You knew my daddy?' And it just broke my heart."

His partner Paul Mackin looks exhausted yet serene. He has another story he wants to tell me.

"My wife and I usually give money to a family every year for Christmas," he says. "Instead of giving Christmas gifts to our kids, we donate $1,500 to a family. A friend of mine told me a story about a 9/11 firefighter whose son was playing in the football championship down in Florida. I found out that he didn't have the money to go. So we made him our family for Christmas."

Mackin and his wife invited the firefighter to their house and handed him $1,500 in cash. He told the man that he should go to Florida with the other parents and watch his son play.

"The man went down to Florida," says Mackin. "His son's team won the championship. Everything was great and wonderful, and another father decided to take the team to Universal Studios." The dad that Mackin helped with some cash said he was too tired and wanted to remain at the hotel so he could relax by the pool.

"So, he goes down by the pool and sees a girl floating face down in the water," says Mackin. "The guy pulls her out of the pool, gives her mouth-to-mouth resuscitation because he's an FDNY firefighter, and saves her. He calls me up from Florida and says, 'You are not going to believe how far your $1,500 just went. I just saved somebody's life.'"

Mackin breaks into a smile as he shakes his head.

"If more people would just pay it forward like that, the world would be a different place," he says.

"This is the place to be. This is where it really happened."

I keep thinking about the last thing Robert McGuire said to me. The retired firefighter I met earlier in the day was explaining why he, Paul Fisher from the Monmouth County Sheriff's Office, and thousands of others keep returning to O'Hara's.

This bar is not where the horror of 9/11 happened. But it's close by. Like the bartender who lets you tell your troubles without needing to know your name, O'Hara's was the place where firefighters could wash the dust and stench of death from their skin as they exited Ground Zero. Like the firefighters who tried to save comrades buried under tons of steel and concrete, O'Hara's was wounded badly—but lived to tell about it.

O'Hara's is not where the horror of 9/11 happened. It is where the hope began.

CHAPTER SIX

Games of Empathy

The elevator opens on the third floor of the Daniels Spectrum, a hub for culture and innovation located in the heart of Toronto's Regent Park neighbourhood. I turn right and walk down a long hallway to a meeting room with the door closed. A middle-aged woman approaches me.

"You're here to meet Sidra." The woman smiles at me. She opens the door and motions me to walk in. Twelve people are seated in a circle in the middle of the room.

"You're just in time," says a man in his mid-twenties who is standing in the middle of the circle. He hands me the weirdest set of goggles I've ever seen: a Samsung Galaxy smartphone turned on its side and fitted into a frame made of heavy-duty cardboard. It's a lot cheaper than the Oculus Rift, a snazzier-looking virtual reality headset built for serious gamers. And less cumbersome. Where these goggles work with a smartphone, the Oculus Rift requires a computer.

Clouds over Sidra is the first ever film shot in virtual reality (VR) for the United Nations. It was shot at the Za'atari refugee camp in Jordan, home to as many as 120,000 Syrian refugees.

Sidra, the 12-year-old girl who stars in the film, has been living at the Za'atari camp since the summer of 2013. The film uses proprietary technology developed especially for Here Be Dragons, a VR production company founded by Chris Milk, an award-winning artist and director of commercials and music videos.

The conflict in Syria began in March 2011. Since the start of the civil war, the United Nations estimates, 400,000 Syrians have been killed, and 6.3 million have been displaced internally. More than 5 million have fled the country, becoming refugees like Sidra.

The young man helps fit the goggles on me. My eyes adjust, and I see a computer display with a list of menu items and a blinking cursor.

"You'll be able to watch the film in a moment," the young man addresses everyone in the room. "Remember, the film is shot in 360 degrees. Try looking up and down, and don't forget to look behind you. There's something to see everywhere you look."

In a TED Talk, Chris Milk, co-director of the film, called VR an "empathy machine." The UN footed the six-figure budget to raise awareness and money for Syrian refugees. My motive in watching the film is to find out how much VR cranks empathy, if at all.

The young man gives final instructions. "It can be very disorienting at first," he says. "Some people get vertigo for the first few minutes. If you're prone to migraines, you might get a headache. Let me know if you experience either."

"I'm glad I didn't eat," says one of my co-participants.

"Okay, press the home key on your phone when you want to begin viewing," says the young man. "We'll get back to you when the film is over."

I push the button. Suddenly, I'm transported to a landscape

of sand dunes and mountains. With my goggles, I can see them all around me. Then a quiet voice fills the stark and barren landscape. "We walked for days crossing the desert into Jordan," says the off-screen voice of a young girl in accented English. "The week we left, my kite got stuck in a tree in our yard. I wonder if it is still there. I want it back."

The desert fades to black, and I am transported into a bedroom inside the Jordanian camp. A young girl is sitting on a bed, facing me. "My name is Sidra, and I am 12 years old," she says.

For the next nine or so minutes, I experience scenes from a day in the life of Sidra. I see her at school and out in the field playing soccer with her friends. I see young men pumping iron in a gym and playing computer games.

I am surprised by the vividness of the scenes that unfold in front of me. Sidra's baby brother shrieks and appears to pass right by me. I turn around and see the toddler walk out the bedroom door.

Near the end of the film, Sidra walks into a big tent where her mother has prepared dinner. The whole family is there. "I still love her food, even if she doesn't have the spices she used to," she says.

Suddenly, I can think of little else except fish sticks and Kraft Dinner—the comfort foods my mother made for me when I was Sidra's age. Unlike the young refugee, my early childhood challenges included learning how to ride a bike and thrive at summer camp without crying myself to sleep from loneliness.

In the final scene, there's a panoramic view of the desert and the cloudy sky. "My teacher says the clouds moving over us also came here from Syria," says Sidra in a voice-over. "Someday, the clouds and me are going to turn around and go back home."

I start to cry. I'm still crying when I hand back the goggles. "Sorry about getting them wet," I tell the young man.

"Don't worry," he says. "We're used to it."

The 12 men and women who watched the film with me are gathered around an energetic man in his late thirties with a dark beard flecked with a tiny bit of grey. Gabo Arora is the first-ever creative director at the United Nations. The artist and filmmaker recruited Sidra for the film and co-directed it with Chris Milk. Arora was second-unit director on U2's first ever VR music video, *Song for Someone*.

He is showing *Clouds over Sidra* to small groups of people like this in Toronto, across Canada, and around the world. The UN is hoping that the virtual experience of being inside a refugee camp and seeing life from the standpoint of a child will help recruit volunteers and raise funds needed to resettle Syrian refugees.

Arora wants you to empathize with Sidra. He thinks that a VR version of Sidra is the way to do it. He and I find a quiet place to chat in a hallway at the Daniels Spectrum, steps from where another group is about to watch the film. Now that I've seen the virtual reality version of Sidra, I want his take on the young girl he worked with in person.

"When you're 12 years old, there's this kind of transition period between being a child and turning into a teenager," says Arora. "You're innocent like a child. You play, and you're trying to be happy. But you're starting to become conscious about what is happening. With Sidra, it was her understanding that this was slowly eroding, that she was understanding that maybe her hope was not possible. Maybe she wasn't going to go home. Yet, she still hung onto it."

Casting Sidra, an attractive and articulate child, was the easy part. The challenge, says Arora, was to find a way to attract empathy for the girl by making her relatable to a diverse audience. "We didn't try to hit strongly on messages like child marriage or the fact that girls should have an education," he says. "I tried to block all of that out of my head. That shouldn't be the aim of this. The aim of it should be for you to just get to know her."

To do that, he looked for the idiosyncrasies of her everyday life, things that aren't dramatic but add layers to her character. He honed in on an attitude Sidra flashed that is common to many girls on the cusp of womanhood. "What was really compelling was her relationship to boys," he says. "There was a sassiness about her and about her relationship with boys. It was like she was saying that boys are dumb. Why are they not studying like she does? Why do they play games that girls don't play?"

Another thing that makes Sidra's personality quirks so compelling is that they seem surprising to people who live in the developed world. "I think we assume in that part of the world, girls of that age are not able to stand up for themselves or to express certain opinions," he says. "She did so, but it was playful instead of disrespectful."

That's not the only contrast that makes Sidra surprising to viewers.

"Another thing about her is that she just goes about her day and her business," says Arora. "She feels compelled to study and play soccer. She is basically living a life as one would in any home of a 12-year-old in many parts of the world. And yet, she is surrounded by barbed wire. It's almost like she can't see any of that."

Arora says the biggest controversy was whether to include a scene near the end in which Sidra cries. "I felt it was maybe the wrong thing to do, especially with a young child," says Arora. "Yet the family felt okay with it. They said that this is something that they don't usually think about. We had a big discussion about the scene with UNICEF. In the end, we thought that it wasn't exploitative. It just felt honest as to who she is."

To Arora, that sort of honesty is what builds the audience's trust of the story and empathy for Sidra. "In my definition of empathy, we want to understand all parts of somebody, good and bad," he says. "If someone is suffering, it's important that you still see them and the truth around them. They are incredibly complicated people who are also laughing, praying, and hoping. You want to see them in ordinary circumstances. There's something about working with people in an ordinary way, especially in virtual reality, that makes you think that you could be just like them."

Empathy for Sidra is the means to a much bigger goal: raising awareness of Syrian refugees, raising dollars for resettlement, and recruiting volunteers. If *Clouds over Sidra* is any indication, it's a home run for philanthropy. A screening for high-impact donors held just prior to the Third International Humanitarian Appeal for Syria in Kuwait in 2015 raised $3.8 billion US—70 percent more money than expected. The UNICEF education team is using the film to showcase the need to support children's education in crisis situations.

The meeting that I attended at the Daniels Spectrum is a pilot version of the Sidra Project, a collaboration between the UN and Artscape, a not-for-profit urban development organization based in Toronto. Funded by 20 organizations that include RBC Royal Bank and the Institute for Canadian Citizenship, the Sidra Project was established "to sustain interest and support for refugees to ensure their successful resettlement in Canada."

The first phase was launched as part of the 2016 Toronto International Film Festival (TIFF). That fall, more than 7,000 people in Canada gathered in groups of between 10 and 60 to watch and discuss *Clouds over Sidra* at schools, private homes, and public venues. According to organizers, 95 percent of those surveyed agreed that the Sidra Project "heightened their sense of empathy toward the plight of refugees." Seventy-three percent of those surveyed said they have taken actions to help with efforts to resettle refugees in Canada, and 94 per cent said they believe the Sidra Project would build more support for refugees if more people experienced it.

"We are now in the process of scaling up the project and sharing what we have learned with others seeking to leverage art for change," said Artscape CEO Tim Jones on the Sidra Project website.

Gabo Arora is also using VR to give viewers a new perspective on some polarizing political issues. Arora and Ari Palitz co-directed the UN-sponsored VR film *My Mother's Wing*, an eight-minute documentary that shows a day in the life of a Palestinian woman living in Gaza. Om Osama is a 37-year-old former school worker who lost two sons when Israel shelled UN Relief and Works Agency school shelters during the summer of 2014.

"The aim of the film is more about peace-building," says Arora. "It's more about trying to get people to understand another community that they don't have a lot of information about."

Arora says he doesn't expect Israelis who see the film to change their beliefs about the conflict in Gaza. But he hopes they'll reflect on the lives of people who live there. "Slowly, you can hopefully see something change," he says. "It's not easy to do a film on Gaza. We want to challenge ourselves artistically. We want to push the medium in a way that will be really good."

* * *

There's doing good in the artistic sense. Then again, there's doing good in the prosocial sense. Filmmaker Gabo Arora is passionate about both. And he's not alone.

There's a plethora of VR projects aimed at the social justice market. In 2015, CNN political commentator Van Jones and television producer Jamie Wong created *Project Empathy*, which tells gripping first-person stories intended to be viewed by legislators and other influence makers. The first series of films in the *Project Empathy* stable take place inside the American prison system. *The Letter*, directed by Wong, features the story of *New York Times* bestselling author Shaka Senghor, who was incarcerated in Michigan for 19 years, seven of which he spent in solitary confinement. One of the lofty aims of *Project Empathy* is to reduce the incarceration rate in the United States by half.

At Stanford University, a study called *Empathy at Scale* puts volunteers in a variety of VR scenes designed to see if they become more empathic to homeless people by experiencing virtual homelessness.

Academics are doing studies to figure out if VR can build empathy. Producers and directors who make films for the viewing public—and the thought leaders who get behind them—already believe it does. "We knew from the feedback we were receiving that the combination of Gabo's artistry and the VR technology was having a powerful effect on people, but the survey results show a level of effectiveness beyond our wildest dreams," enthuses Artscape CEO Tim Jones on the Sidra Project website.

Film directors have been making documentaries about worthy subjects for decades. Arora says VR is different. "We wanted to bring people to places they can't go with any regularity. There's no other medium that I can think of that gives me that kind of thing."

The filmmaker says a lot of his favourite non-VR documentaries are considered by others to be slow and boring. "There's something about VR that is a lot more engaging and a lot less boring because you're exploring a space."

Arora's collaborator Chris Milk is even more effusive in his praise of this new medium. In 2015, Milk did a TED Talk entitled "How Virtual Reality Can Create the Ultimate Empathy Machine."

"It connects humans to other humans in a profound way that I've never seen before in any other form of media," says Milk. "And it can change people's perception of each other. And that's how I think virtual reality has the potential to change the world. So, it's a machine, but through this machine we become more compassionate, we become more empathetic, and we become more connected. And ultimately, we become more human."

Paul Bloom, professor of psychology at Yale University and author of the 2016 book *Against Empathy: The Case for Rational Compassion*, believes that VR is engaging and may even help to combat indifference to the suffering of others. That said, in an article in *The Atlantic*, Bloom argues that "VR doesn't actually help you appreciate what it's like to be a refugee, homeless or disabled."

Moreover, he says a virtual walkabout through the Za'atari refugee camp where Sidra lives is misleading. "The awfulness of the refugee experience isn't about the sights and sounds of a refugee camp; it has more to do with the fear and anxiety of having to escape your country and relocate yourself in a strange land," writes Bloom. "Homeless people are often physically ill, sometimes mentally ill, with real anxieties about their future. You can't tap into that feeling by putting a helmet on your head."

Bloom makes an excellent point. Though her narration provides some clues, spending a few minutes looking around the Za'atari refugee camp doesn't tell me what it's like to *be* Sidra. I can't even be sure that I was shown an accurate reflection of what Sidra sees on a typical day. For all I know, there could have been dangers or even delights just beyond the frame that I was being shown.

Arora thinks Chris Milk's TED Talk comment about VR as the "ultimate empathy machine" has been misinterpreted. "VR *storytelling* is the empathy machine," he says. "VR must be combined with the crafting of the filmmaker. I think that's underestimated at this point because people are so wowed by the technology."

We should also not forget that the purpose of the exercise is to get viewers to empathize with Sidra, not become Sidra. Unless you're a mindreader, empathizing with someone else does not mean being able to crawl inside that person's head and know what they're thinking.

Still, I don't find myself wondering what Sidra is up to these days. I have little or no stake in what happens to her and her family. Aside from the startling experience of virtual reality, the story itself is both familiar and unsurprising. It made me cry momentarily, but it didn't change me. But as a fundraising tool, I think *Clouds over Sidra* is a winner. Getting donors to contribute is a worthwhile goal that makes the world a better place. But after you cut the cheque, there's a good chance that a film like this accomplishes little more.

So far, I find myself thinking that VR should be able to do something more substantial in building empathy. It turns out that some researchers agree.

* * *

Virtual reality has become a hot tool for neuroscientists who study empathy. Like Gabo Arora, who uses VR to transport viewers to a different part of the world, these researchers are running experiments with VR to transport human subjects to a digitally generated place. The technology can put people with a deathly fear of flying into a virtual cabin at cruising altitude, or take those with arachnophobia into a virtual pit of tarantulas. VR is also a promising therapeutic tool for first responders with post-traumatic stress disorder (PTSD).

Mel Slater, a research professor at the University of Barcelona, is using VR in an intriguingly different way, one that has implications for the study and manipulation of empathy. He uses the technology to turn adult human subjects into kids.

Slater and his colleagues have recruited volunteers who don an OptiTrack full-body motion-capture suit and head-mounted display. The suit is exactly like the ones actors use so that they appear on film doing everything from performing death-defying stunts to being transformed into talking apes and other-world aliens.

In Slater's experiments, the subjects aren't movie actors performing stunts but ordinary humans performing mundane tasks, which the OptiTrack suit captures faithfully, sending the data to a computer. The computer then creates an avatar that the subject can see on a head-mounted display.

That's where Slater's experiment gets interesting. The computer is programmed to pull a fast one on the research subject, scaling his or her size down to that of a four-year-old child. When the subject holds his or her own physical hands up, the head-mounted display shows the virtual hands of a child performing the same action. When the subject looks into a virtual mirror, he or she sees the reflection of a child.

Slater and colleagues recruited 30 adults and turned them into virtual four-year-olds. Within seconds, the adults were so transformed by the illusion that they overestimated the size of objects, just like a preschooler. As part of the experiment, participants were shown a virtual room filled with adult-sized furniture and a room with pieces sized for small kids. The subjects estimated sizes and distances more accurately in the virtual kids' bedroom than in the one for virtual grown-ups.

Hospitals are scary places for kids, and that is something adult doctors like me often fail to appreciate. We are often admonished to "see things from a child's point of view." This technology is a pretty easy way to do exactly that.

The bottom line is that the researchers were able to flip a switch in a grown-up's brain that gave them the perspective of a child. The next experiment by Slater was designed to see if he could do the same with skin colour.

Humans are hard-wired to prefer those who share their skin colour. If you meet a stranger with your same skin colour, you see that person as likeable, trustworthy, and sympathetic. If the stranger has a different skin colour, you're more likely to think he or she is pushy, devious, and mean. You can't help yourself; your brain makes those judgments automatically in about as much time as it takes to identify the stranger's skin colour.

Psychologists have discovered a pattern in human behaviour. If you think a stranger is a member of your racial group you unconsciously mimic their behaviour. If the stranger crosses his legs, so do you. If the stranger uses inflections like uptalk or vocal fry when speaking, so do you. On the other hand, if you encounter a stranger whom you perceive to be a member of a different racial group, you're far less likely to mirror that person, and far less likely to form a social bond. In a paper published in

1999, psychology professor Tanya Chartrand, then at Ohio State University, wrote that mimicry occurs automatically the instant you perceive that a stranger is like you. She called the tendency to mirror people the *chameleon effect*.

In his skin colour experiment, Mel Slater used the same set-up that turned adult research subjects into virtual four-year-olds. But this time, instead of changing the age of the subjects, Slater changed their pigmentation. The subjects (all Caucasian) donned the OptiTrack suit and head-mounted display. Their movements were sent to a computer, which generated an avatar of themselves that the subject could see on the head-mounted display. Half of the subjects were given a Caucasian avatar, and the other half were given one that was black. In the experiment, a second virtual character appears on the computer screen, and the experiment calls for each subject's avatar to interact with the second character. Meanwhile, researchers measured how much the subject (through the avatar) mirrored the movements of the second character. Half the time, the second virtual character was Caucasian, and half the time the second character was black.

Not surprisingly, the Caucasian subjects who were given a Caucasian avatar mimicked the second character if that character was Caucasian, but not if the second character was black.

Astonishingly, when these Caucasian research subjects were given an avatar that was black, they mimicked the second character *only if that character was black*, not Caucasian. In other words, whether the subjects liked and mimicked the second character depended not on the subject's own skin colour but on the colour of the avatar. At the flick of a virtual switch, the researchers induced the subjects to turn Other into Self with surprising ease.

"Such virtual race transformations may be an effective

strategy for combating automatic expressions of racial bias," the authors wrote in 2017 in the journal *PLOS One*.

Stop and think about that for a moment. Slater has demonstrated that a quick trip into virtual reality can trigger at least a temporary reversal in racial bias. How long does it last? Can it be made permanent? Does it work for everyone?

The experiment by Slater has caught the attention of a team of young scholars in Winnipeg who are working alongside Survivors of Canada's Indian residential schools, the now abandoned and much-condemned boarding schools funded for decades by the Canadian government. The aim of the researchers is to foster empathy for Survivors to promote and deepen reconciliation between Indigenous and non-Indigenous people using twenty-first-century technology to take people inside a virtual replica of an Indian residential school.

"I hope it doesn't feel too snug," says Dylan Fries, director of interactive media at Electric Monk Media in downtown Winnipeg, as he adjusts the straps of the Oculus Rift over my eyes.

The Oculus Rift headset displays images shot in 3D and has rotational and positional tracking. That means that when you move your head, the field of vision in front of you changes to simulate what the view in front of you would look like if you turned your head in the real world. The headset also comes with integrated headphones that provide surround sound.

Dylan Fries isn't a psychologist studying empathy, or an academic trying to promote reconciliation between Indigenous and non-Indigenous people. Fries builds computer games. His company, Electric Monk Media, is a creative hub that makes everything from VR interactive games to immersive horror films.

Developers are building VR games for Oculus Rift and other competitive devices to gain a share of a worldwide market expected to rise to nearly $23 billion US by the end of 2020. That's out of a total market for augmented and virtual reality of $143.3 billion. That's where the investment dollars are going. It also just happens to be the best place to look for someone with the skills necessary to build a VR version of an Indian residential school.

"This game is called *Phantom of the Forest*," says Fries, pushing a few buttons on a nearby computer to load the program.

Phantom of the Forest is a VR game made for families. After I learn how to operate the controls, the demo begins. Suddenly, I'm a great grey owl in a boreal forest, flying high above the trees on a brilliantly sunny winter's day. For fun, I enter a deep dive and head straight toward a tree. "I flew through the tree, and the snow came flying off," I exclaim. "That was amazing!"

A man in his forties seated nearby takes note of the sudden change in my demeanour. Struan Sinclair is the project leader of *Embodying Empathy: Fostering Historical Knowledge and Caring through a Virtual Indian Residential School*. That's the name of the project to create a VR version of an Indian residential school. Sinclair is also the director of the Department of English, Film, and Theatre Media Lab at the University of Manitoba.

"The tone of your voice sounds different. It was almost wonderment," says Sinclair.

"I was surprised," I tell him. "I feel like I made something happen."

Sinclair is measuring my reaction to *Phantom of the Forest* to see how emotionally engaged I am. He says the success of *Embodying Empathy* depends on the program's ability to establish an emotional bond with participants.

Sinclair comes to the project not just as an academic but as a writer, and a very good one. His 2009 debut novel *Automatic World* was a multiple awards finalist. He is collaborating with Fries on *Tomorrowless*, Sinclair's first interactive VR graphic novel.

"We're going to show you a demo of *Embodying Empathy*," says Sinclair. "It's a prototype. Dylan did the modelling and programming for it. It's just two rooms."

The program loads, and I find myself standing on a dirt road facing a two-storey building made of brick, with a gabled roof. It's the exterior of a typical Indian residential school that existed in the early twentieth century. Church bells toll in the background, and birds are chirping. The Oculus Rift headset gives me a full 360-degree view. To my left is a small vegetable garden and beyond that a wooded area that makes the school seem cut off from the rest of the world. To my right, there is a wrought iron fence painted black and what look like fruit trees. Behind me is a dirt driveway. Looking up, I see dark clouds in the sky, with streaks of sunlight poking through. It's morning, the time of day when many residential school Survivors say they were first brought to the school.

Navigating with the help of a computer keyboard, I make my way up the concrete steps that lead to the front doors. As I walk in, I try to imagine being a child as young as five, leaving my family and walking into this place for the first time.

I look behind me, and the front doors close almost immediately.

On the main floor to my left are two rows of benches and work tables with sewing machines. The area is empty. Turning to my right, I see a staircase that leads to the second floor. On my way there, I stop at a wooden table with a chalk slate on which visitors can write a message.

The turns make me feel slightly nauseated.

Upstairs, there's a dormitory with 20 cots. Halfway down the room, I find a small table with a tape recorder that starts to play as I approach the machine. "On my first day in boarding school, my dad delivered my brother, older sister, and me to this huge, cold, cucumber-smelling place that chilly September day, in 1949 or 1950," says the voice of a young woman. "I remember the priests and a couple of nuns pulling us apart as we desperately tried to cling to one another. I swear those nuns looked like identical twins."

The anecdote goes on for another minute.

As I make my way out of the school, I still feel a bit queasy from the VR experience. I'm not surprised to feel a bit sad for the young woman who spoke about the day she was separated from her siblings. I *am* surprised as I begin to feel something both personal and unexpected. As an eight-year-old in grade 2, I got the strap after talking back to my teacher. The details aren't important, and they don't equate in any way with what kids who lived at residential schools experienced. What is important is that a brief virtual immersion in a facsimile of a residential school made me recall one of the most humiliating times in my entire life.

I tell Struan Sinclair what I experienced, and the brief but powerful empathic connection I made with Survivors. He isn't surprised.

"The demo is very brief but it cues memories," he says. "It is a striking example of the kinds of things that the technology, well-engineered and imagined, can supply."

Sinclair says the finished product will be even more powerful. "There will also be plenty of moments where you'll be hearing testimony, you'll be interacting, you'll be listening, you'll be paying attention. One of the things you want to be conscious of when you design a VR world like this are those attention points where you want people to stop, think, reflect, and look a little harder into something."

If I had seen the same artifacts in a museum display, I'm not sure I would have reacted in the same visceral way. I ask him why the VR experience is so much more compelling than a museum exhibit.

"It's because we've turned it into a game," says Sinclair matter-of-factly, letting the thought sink in.

If you google the words *game* and *empathy*, you get many hits. *Who Am I?* is a game that helps kids think about different ways people self-identify their race and cultural identity, and it gives adults tips for talking about diversity with kids. *Cool School* teaches kids conflict resolution skills by watching animations of realistic situations and seeing the effects of good and bad choices. *Middle School Confidential* is a graphic novel app that helps kids learn to identify emotions, reflect on personal strengths and weaknesses, respect the viewpoints of peers, and build friendships.

"As a game designer, you are creating situations for players to experience and then letting them experience it," says Dylan Fries. "They have their own layer of emotions and reactions to the game, and empathy is a big part of it. With VR, players feel more placed in the world and spend a lot more time looking at details and talking about how the environment makes them feel."

I ask Fries if his take on empathy is unique within the game industry.

"I think it's something that often gets left out of games," he says. "They're aiming to make them fun, so they skip over consequences. You just die and respawn, and *boom!* you're a new person. But they are moving toward making more games that tell non-violent stories, and ones that explore a range of emotions and experiences."

Fries credits part of that shift to the growing influence of filmmakers.

"We have film people on our team who are used to dealing with empathetic situations and a range of emotions that are much wider than you normally deal with in a traditional video game," he says. "Compared to a film, I think a VR game can be more powerful at providing empathy. In a movie, you're watching somebody else go through an experience. In VR, we put you in the situation, so you have some agency to make a decision and then see the consequences."

Sinclair agrees. "I think we're at the beginning of a turn in VR gaming toward much more complex environments," he says. "What Dylan and his crew and others are doing is new, but it's also really exciting in terms of its potential to go beyond games for entertainment."

Going "beyond games for entertainment" means creating games for serious purposes. *Virtual Heroes* builds 3D simulations for medical and military training. *Gamelearn* teaches office workers everything from time management to conflict resolution. Game theorists and academics call these "serious games." For Canadians, none is more serious than *Embodying Empathy*.

Sinclair says it's time to leave the gaming studio and meet an Indigenous elder and the others on the project.

"Can you tell me what you had for breakfast so I can make sure I've got a good level on the recorder?" I ask Theodore Fontaine. He is a 75-year-old former chief of the Sagkeeng First Nation in

Manitoba. The breakfast question is my stock method of trying to put the people I interview at ease.

"I had muskrat for breakfast along with bannock," he replies.

Fontaine is an Indian residential school Survivor. When he says bannock, my antennae go up. *Bannock* is a Scottish word for a kind of flat, round quick bread. It's a staple dish for most Indigenous nations across North America. In his 2011 memoir, *Broken Circle: The Dark Legacy of Indian Residential Schools*, Fontaine describes in detail the bannock his mother made for him and his brothers and sisters when they were kids, and how that indelible memory stood in stark contrast to his life at an Indian residential school.

I'm trying to build rapport with the Indigenous elder and Survivor of 12 years in a residential school by reminding Fontaine how lovingly he writes about the comfort food his mother made for him when he was a small child.

But Fontaine is pulling my leg. "I didn't actually have bannock and muskrat this morning," he says with a grin, pleased that I've fallen for a little teasing.

He's using humour to call attention to some old assumptions he's working to change. Where I might see myself as a third-generation Canadian Jewish man, Indigenous people might see me as a privileged white settler.

Thomas King writes about this perspective in his 2012 book *The Inconvenient Indian: A Curious Account of Native People in North America*: "Native history in North America as writ has never really been about Native people. It's been about Whites and their needs and desires. What Native peoples wanted has never been a vital concern, has never been a political or social priority."

Settler colonialism is a revisionist narrative in which colonists replaced Indigenous peoples. Framed that way, correcting the historical record means digging into the histories of Indigenous people and finding out how that replacement occurred. Accomplishing that requires that settlers like me imagine what it was and is like to be an Indigenous person in Canada. If anyone can do that, it's Theodore Fontaine.

The Indigenous elder is a formidable man. At 32, he graduated as a civil engineer and led a mineral exploration crew in the Northwest Territories. He was made chief of the Sagkeeng First Nation before the age of 40. After that, he had stints working for the federal government. He was executive director of the Assembly of Manitoba Chiefs and strategic advisor to the chiefs on Indian residential school issues.

From reading his book, I gather that Fontaine takes immense comfort in remembering how blissful his life was before it was shattered by the people who ran the school. Indian residential schools did not arrive by accident. In the nineteenth century, the Canadian government was consumed with the acquisition and settlement of territory as it built the nation. Authorities saw Indigenous groups as an obstacle to that process and referred to their very existence as the "Indian problem." They saw assimilation of Indigenous people as the solution. The most articulate and outspoken advocate of that approach was Duncan Campbell Scott, who served as deputy superintendent of the Department of Indian Affairs from 1913 to 1932. "I want to get rid of the Indian problem," he wrote in defence of the *Gradual Civilization Act*, a bill passed by the 5th Parliament of the Province of Canada in 1857. "I do not think as a matter of fact, that the country ought to continuously protect a class of people who are able to stand alone . . . Our objective is to continue until there is not a single Indian in Canada that has not been absorbed into the body politic and there is no Indian question, and no Indian Department, that is the whole object of this Bill."

The federal government believed that Indigenous children could be assimilated far more easily than adults. It embraced a model known as "industrial education," which involved forcibly removing Indigenous children from their parents and from their communities and placing them in Indian residential schools. The assumptions were that Indigenous people could not adapt to modern society without assimilating, and that the only path to successful assimilation was to adopt Christianity and to speak English or French exclusively. Upon arriving at the schools, children were stripped of their Indigenous names. The clothing provided by their parents was taken away and destroyed. Indigenous children were discouraged from speaking their Native language or practising Native traditions, and punished if caught trying to do so. Most of the children went without seeing their parents for 10 months of the year or longer. An estimated 150,000 First Nations, Métis, and Inuit children attended the schools, the last of which closed in 1996.

On a warm day in September 1948, just days after his seventh birthday, Theodore Fontaine's parents brought him to the Fort Alexander Indian Residential School. It was a couple of kilometres on foot from Fontaine's childhood home at Treaty Point on the Winnipeg River, but it may as well have been on a different planet.

In his memoir, Fontaine writes with aching pain and bitterness about the day his parents left him at the school and his happy childhood was shattered: "Little did they know that the experience I was about to undergo for the next 12 years would shape and control my life for the next 40 or 50. From this point on, my life would not be my own. I would no longer be a son with a family structure. I would be parented by people who'd never known the joy of parenthood and in some cases, hadn't been parented themselves."

Fontaine experienced all manner of cruelty. Like many of the thousands of Indigenous children who lived at residential schools, he experienced sexual abuse. "I could talk about physical and sexual abuse, and mental and spiritual abuse," Fontaine tells me. "I've had to deal with that aspect for years and years and years."

In the first chapter of his book, he writes about H., a shop teacher who sat Fontaine in front of him and commanded the boy to pull the teacher's exposed penis "out by its roots." Fontaine also describes the ritual exercise known as *ménage*, in which he and his classmates would be selected to have their prepubescent genitals washed by Father P. "That's just a minor description of being abused by an adult," Fontaine tells me. "You can imagine other ones that I haven't mentioned. Some of my presentations to the public get graphic, but they're not in the book."

Thirty-two hundred children are documented to have died at residential schools, but the actual number is believed to be much higher. According to the Truth and Reconciliation Commission, the number one identified cause of death was tuberculosis. Other causes included influenza, pneumonia, and lung disease. In nearly half of the death records of named and unnamed children who died at residential schools, the cause is not identified.

Fontaine spent 12 years at residential schools, the first ten at Fort Alexander and the last two at the Assiniboia Indian Residential School. He emerged from the experience severely traumatized. Charles Brasfield, a psychiatrist in British Columbia, has coined the term *residential school syndrome* to describe the symptoms experienced by Survivors; these include "recurrent intrusive memories, nightmares, occasional flashbacks, and quite striking avoidance of anything that might be reminiscent of the Indian residential school experience."

Fontaine says it took close to a decade for him to begin to process what happened.

"It goes back to the late '60s, early '70s, when I was suffering from the syndrome," Fontaine recalls. "I started learning that these horrors that came up were persistent. Each time they would come back, I'd sit in the quiet or by the river or in the forest and drink myself silly."

Eventually, he figured it was better to expect flashbacks than to push or drink them away. "It became a little easier to anticipate what was going to come," he says. "It was things that were constantly invading my mind and my life."

Each flashback led to more recovered memories and insights that he kept on scraps of paper. But it wasn't until 2010 that Fontaine sat down to write *Broken Circle*. That's when he realized what it was about the school experience that traumatized him the most. Not the physical abuse. Not even the sexual abuse. Something more elemental.

"I think it was the whole process of destroying who I was, what they called the 'little savage boy,'" Fontaine tells me in a voice both sad and angry. "The whole prospect of trying to destroy our culture, our languages, and our own people . . . to the point that we're called savages."

In absorbing that core message of the residential school, he learned with stunning speed to loathe his people and himself. "I had entered the school in 1948, and my first summer holiday was in 1949," he says. "When I went home, I can honestly say that I hated my parents, my grandparents, and my extended family that was part of the reserve because they were Indians. I think that is one of the hardest things I live with right now."

What Fontaine and thousands of other children experienced is cultural annihilation. An article co-written by *Embodying Empathy* project leader Struan Sinclair and his colleagues describes the lasting impact on children losing their Indigenous cultural identity: "Children also lost out on much of their cultural learning, as entire story cycles were missed while students boarded at school over the winter. Songs, practices, traditions, games, jokes, ceremonies and other aspects of Indigenous cultural learning were not transmitted. Upon leaving school, students often lacked the cultural resources required to maintain a feel for Indigenous group life."

The Indian Residential Schools Settlement Agreement (IRSSA), approved in 2007 by the Supreme Court of Canada, provided $1.9 billion in financial compensation to attendees of residential schools. The Aboriginal Health Foundation received $125 million. Sixty million dollars was earmarked to create the Truth and Reconciliation Commission (TRC), established in 2008 to document and publicize the extent and impact of the experiences of Indigenous children at residential schools, and to provide a safe place for Survivors to tell their stories. The TRC issued its final report in December 2015. The cultural, physical, and sexual violence perpetrated on Indigenous children caused all manner of suffering, much of which was documented by the TRC in the testimony of Survivors.

The *Embodying Empathy* project is based in Winnipeg. That's a good fit, since Winnipeg is also home to the National Centre for Truth and Reconciliation, which holds 7,000 statements from Indigenous Survivors, plus photos, videos, and millions of government documents. The Canadian Museum for Human Rights is also located in the city.

The *Embodying Empathy* project is funded by a grant from

the Social Sciences and Humanities Research Council (SSHRC), a Canadian federal research-funding agency that supports research and training in the humanities and social sciences, with a special emphasis on Aboriginal research. In an article, Sinclair and colleagues described the project as "a critical and creative collaboration between scholars, IRS Survivors, and technologists who are working to construct a virtually immersive IRS 'storyworld'" with the aim to "alleviate suffering and further reconciliation—not just between Indigenous and non-Indigenous Canadians but between IRS Survivors, their families, and their communities—through the cultivation of a morally and politically salient form of empathy."

Those colleagues, all from the University of Manitoba, include Adam Muller, who is working on ways to represent digitally the sensitive subjects of genocide and mass violence; Andrew Woolford, an internationally recognized expert in genocide studies and settler colonialism; and Greg Bock, an expert on technical and ethical issues related to the digital archiving of records.

The aim of the project is to use testimony and records assembled by the TRC to immerse visitors in a virtual residential school. Nothing like this has been attempted anywhere else, in part because of the scale, and in part because the project's leaders include Survivors of Indian residential schools.

Although they have 7,000 testimonies to draw upon, Sinclair and colleagues have recruited a small group of six Survivors, including Theodore Fontaine, as Andrew Woolford explains.

"Theodore Fontaine, who has written his memoir *Broken Circle*, which is fabulous, was our key liaison and key champion of this project and brought it forward," says Woolford.

Fontaine sees the project as a natural extension of his own story and that of Indigenous people across Canada. "*Embodying Empathy* wasn't created out of the blue," says Fontaine. "It came as part of reconciliation, where (bless their souls) people like Professor Woolford were taken aback by what residential schools were all about. He wanted to do something so that there should be some understanding and human feelings toward what really happened and that it not happen [again]."

Woolford says that much of the early work with Fontaine and fellow Survivors was about building trust. A big part of that meant putting Survivors like Theodore Fontaine in control of the process and of the outcome. They, not academics like Woolford, are the experts, the ones with knowledge and experience of what happened at Indian residential schools.

Getting Canadians to make an empathic connection to residential school Survivors is one worthy social cause. But these days, it's not the only one. In the latter half of 2017, Hurricanes Harvey, Irma, and Maria caused storms and floods that left a trail of destruction, death, and illness in their wake in Texas, Florida, and the Caribbean. We also witnessed the aftermath of an earthquake in Mexico. Then, there are tsunamis in the Pacific, and wildfires in Alberta and other parts of western Canada. Climate scientists say we can expect many more episodes of extreme weather in the years ahead.

There is no end to the demands made on the public, amplified by compelling photos and mass and social media coverage, to care about one cause after another. Academics say this surfeit of tragic news leads to moral numbness, what they refer to as *empathy fatigue.*

Enter VR games. It's a trendy idea among scientists that you get people to care about Survivors of residential schools and other causes by getting them to play a game.

The immersive experience that Struan Sinclair, Adam Muller, Andrew Woolford, and Greg Bock want visitors to *Embodying Empathy* to absorb is that of being a child in an Indian residential school, complete with many of the discomforts, including the pang of separation from parents and the fear and confusion of entering an environment that is utterly alien. For ethical reasons, they have decided against depicting scenes of sexual abuse.

Given the need for the active participation of Survivors, there is a risk of retraumatization. VR games have reportedly triggered flashbacks in people with post-traumatic stress disorder. Andrew Woolford says that seemingly innocuous things can trigger flashbacks: "So many things can be a trigger, from the smell of a certain sort of floor wax to a sound. It's very hard to protect against that. So, how are we going to prepare people for entering this world?"

Fontaine says they've taken special precautions. "We've had healers sit with us in case there's something that triggers some awful memory. It still can happen, but hopefully, people are very careful in seeing this."

I want to know if Fontaine and other Survivors have put on the Oculus Rift headset.

"I am anxious to see the final project, but it's not a priority for me," he says.

In addition to building *Embodying Empathy*, Sinclair and his colleagues plan on studying its impact on visitors. He says they have designed *Embodying Empathy* so that they can keep track of how visitors respond to various parts of the program, and where they spend most of their time.

Fontaine is convinced that VR will build empathy for Indigenous Canadians. "The Creator looks after stuff," says

Fontaine. "As a nation, we're very lucky to begin to grasp virtual reality. Sometimes, technology is scary, and sometimes it has its use. I think we can't let this drop. We have to make sure our people survive, and one way of doing it is through technology."

I decided to tell Theodore Fontaine that when I watched the *Embodying Empathy* demo, I got a flashback to the day I got the strap. As I told him the story, I was conscious of how trivial my childhood humiliation was compared to what Survivors like him experienced.

Still, he's not just a Survivor but also the Indigenous elder on the team that has built *Embodying Empathy* to emotionally engage people like me. I needed to know what he thought of my free association. Far from being self-conscious, Fontaine heard my story with respect and then sought to forge a connection with me.

"I want to say something about the strap," he says. "You're not an Indian person, but the strap you experienced was probably the same kind of strap that we had, you know. They were made from the old tractors that had these huge straps that made the engine go."

Fontaine uses my story to reflect on his own experience and connect it to mine.

"People including teachers at the school carried a six- or nine-inch strap in the pockets of their robes," adds Fontaine. "Those were used commonly not only in the residential schools, but in schools where kids like you received it. It's traumatic, you know. You must have been pretty young."

"I was eight," I reply.

"A lot of people don't have an experience that lets them feel what happened to that little eight-year-old," he says. "What you experienced is the main purpose of this project."

Fontaine's capacity to empathize with me is both astonishing and humbling. He thinks my highly personal reaction to the *Embodying Empathy* demo of an Indian residential school helped forge a connection to his suffering personally—taking it from my head and giving it to my heart.

CHAPTER SEVEN

Homeless in Brazil

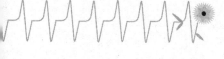

The Brazilian metropolis of São Paulo is so sprawling it takes close to an hour just to get from the airport to the centre of town. I pass clusters of skyscrapers and linked residential neighbour-hoods along a vertigo-inducing pattern of freeways that extends to the horizon in every direction. This is the largest metropolis in South America; with a total population of 12 million, it's a megacity.

In the downtown core, every street corner bursts with new businesses, trendy bars, coffee shops, and burger joints. At night, the same streets fill with young men and women dressed in stylish clothing, flirting with one another.

That São Paulo is also Brazil's wealthiest city is evident. But there's an undercurrent of tension. Look closer, and you see stores that have gone out of business and entire buildings that sit empty. Brazil is mired in a recession that began in 2010 and shows few signs of abating. Unemployment and bankrupt-cies remain high. A corruption scandal in 2015 that led to the impeachment of President Dilma Rousseff a year later—a scan-dal that now involves current president Michel Temer—continues to erode trust in the government.

São Paulo is a place where there's tension between the haves and have-nots. As my jovial taxi driver remarks on the drive in from the airport, "This is a place where you can get lunch for 15 *reais* or 150 *reais*. There's something for everyone." These juxtapositions are often quite visibly on display—informal *favela* communities, or shantytowns, abut luxury high-rises, while groups of homeless people sit alongside the most important financial headquarters on the continent.

Rich and poor live in very different parts of the city. But sometimes, their lives intersect in life-altering ways.

On a brilliantly sunny Monday in April, a luminous woman named Shalla Monteiro emerges from her condominium. I'm in Vila Madalena, a quiet upper-middle-class neighbourhood Shalla moved to a couple of years ago. Located near the University of São Paulo, it's a bohemian part of the city being transformed through gentrification—bustling nightlife, charming cafés, colourful graffiti-filled alleyways, ritzy art galleries, and yoga studios.

Shalla is in her early thirties, with piercing eyes that draw you in when she talks. She inspires instant trust: reassuring you when she speaks, taking your arm when crossing the street. There's an aura about her. I like her the instant I meet her.

I'm here to meet Shalla because I've been told her empathic ability is unworldly. I've come all the way to Vila Madalena to learn where that gift for kindness comes from. To do that, I must first get used to the fact that Shalla is spontaneous and always open and ready to make a human connection. Like right now. Behind Shalla, a three-year-old boy scoots out of the elevator on a wooden pedal-less bicycle.

"Nice bike," I comment.

"It's a motorcycle." The child offers a prompt correction before cruising around the lobby, his red Velcro sneakers propelling him forward, and then suddenly braking. Lautaro is Shalla's first-born son, Tata for short.

"Let's go see Uncle Adriano?" Shalla asks him.

"*Vamos!*" Tata exclaims, zooming ahead of us on the pedalless moto, masterfully manoeuvring down little ramps, and braking a little too close for comfort at the sidewalk's edge.

Up ahead, we see Adriano standing at the corner, surveying the street. From a distance, he looks like a bouncer in a nightclub, arms folded across his chest, blue sunglasses staring coldly out of a grey checkered hoodie. Seeing Tata and Shalla, his posture softens and a broad smile spreads across his face. He comes across to greet us.

"Are you going for a walk?" he asks, beaming down at Tata. "Where are you going?"

"To the park!" Tata responds with delight. Adriano holds the top of Tata's head and gently steers him across the street, looking out for cars.

There is a faint smell of alcohol on Adriano's breath, and I notice that he hasn't removed his sunglasses to speak with us. I ask him about Tata.

"Ah, he's my little friend," he says, smiling at the child, who is absorbed in an imaginary world of elaborate motorcycle adventures. "We just started to talk, then we started to play together. We joke around. We just became friends." He repeats the last sentence several times.

Adriano doesn't quite have the words to describe their relationship, but their body language reflects an exceptional level of physical closeness and intimacy.

Suddenly, a group of men pass by in a car. Adriano recognizes them. He greets them and bums a cigarette. Shalla gives him a disapproving look but says nothing. Tata watches this.

"I like making friends, meeting new people," Adriano says. "My life is often judged by others." The cigarette balanced between his lips makes Adriano even harder to understand. "Some people don't like me, there are prejudices," he trails off, mumbling. "I like it here, it's a beautiful neighbourhood."

Tata has started tugging on Shalla. "He wants to talk with you, Adriano," she says.

"Go ahead," Adriano says.

"I don't like smoking!" Tata says in a commanding voice. It's fascinating to see the child admonish a grown-up.

"No?" Adriano says, smiling. It's one of those toddler moments when everyone tries to keep a straight face on a serious topic.

"Because smoking is bad—it's bad for you," Tata says, a bit softer.

"Yes, that's true. I know, *gatão.*" Adriano calls him "cat," squatting down to Tata's level and holding him by the shoulders. "Say to me, 'You cannot smoke.'"

Adriano wraps the child in a bear hug over the motorcycle and begins to steer him. Tata holds the top of his hands and squeals in pure delight as they drive around. "Look at this motorcycle driver!" Adriano exclaims.

Adriano is both well known and invisible. The homeless man has been a fixture at that same corner for many years. People walking their dogs, or wearing suits, or pushing strollers all seem to know him, perhaps not by name, but they greet him as part of their regular routine. But almost no one actually *sees* him, except for a little boy with an unusual gift for empathy, and his mother.

* * *

Homelessness exists throughout much of the world. It's complicated by the intersecting challenges of drug addiction, psychological trauma, migration, and domestic violence. In the ER where I have worked for more than 30 years, not a shift goes by that I don't see at least one homeless person, especially in winter. Health professionals see them as wanting something from us, a sandwich, a used pair of shoes, or shelter from the cold. I don't think I've ever had an exchange as long and as judgment-free as the one Tata just had with Adriano.

In São Paulo, homelessness is as massive as it is inescapable. A severe affordable-housing shortage has led to the establishment of large, informal settlements and squatter occupations. A staggering 16,000 homeless people live on the streets of São Paulo, or in temporary shelters, according to the 2011 government census. Some say it's closer to 21,000. The growth in São Paulo's homeless population far outpaces the city's population growth.

It feels as if there is a homeless person on every street corner. On a busy thoroughfare in São Paulo's centre, throngs of people step over and around a man sleeping in the middle of the sidewalk, tattered clothes exposing his skinny frame. Nearby, some people rummage through trash, looking for food or soda cans to collect and sell. Occasionally, someone stops briefly to give spare change to the folks perched along the side of the avenue. Outside a supermarket, a homeless mother holds a six-month-old baby. "Please, can you buy me some diapers?" she pleads. But for the most part, homeless people are treated as part of the cityscape, like graffiti and street signs.

The sight of homeless people produces an array of emotions. Guilt. Disgust. Pity. But also fear. A wrong turn near a city government building takes me beneath a dark overpass, where a group of men are partially hidden behind an upright mattress, lighters briefly illuminating their faces. I nervously cross to the other side of the street and quicken my pace toward the brightly illuminated condominium just beyond the edge of the overpass.

Most people have brief one-off interactions with homeless people, leaving pocket change, leftover food, and other guilt offerings. Establish an ongoing relationship? Forget it. Mostly, we try to get the painful images of how these people live out of our minds as we pass by.

That is what makes Tata's connection to Adriano amazing. Shalla explains that she first noticed Adriano when she, her husband, Ignacio Garcia, and Tata moved into the neighbourhood. "I realized that he was homeless, and that he was a person with an internal restlessness," she says.

I note her use of the phrase "internal restlessness." It suggests she's trying to see things from Adriano's point of view.

Shalla says that she passed the homeless man several times, attempting to make eye contact. But he never returned her gaze. "And then, one day I was with Tata," she recalls. "And there was Adriano sitting on the edge of a flower bed in front of a building where he often hangs out."

Tata looked at Adriano, turned to Shalla, and said, "Mom, that man is sad."

At that moment, Adriano was talking to himself, gesturing and experiencing what Shalla describes as a "very turbulent moment." In that moment, many parents might have labelled the man as disturbed and redirected their child's attention toward someone less "turbulent." But Shalla is not at all like other parents.

"Listen, Tata," she said. Shalla's voice has the texture of syrup and she speaks in a lilting way. "We're going to approach him slowly, we're going to get close, and who knows? Maybe he will feel better, talking to us. Maybe not. Let's see."

As they approached Adriano, he looked up. Sitting on the ground, he was exactly Tata's height. "When Tata and Adriano looked at each other," Shalla says, "when their gazes met one another . . . Adriano's expression changed."

I get chills hearing Shalla describe her young child having this profound moment of connection.

"It was instantaneous," the young mother recalls. "With the child's gaze, he relaxed."

Shalla bent down so that all three were sitting at the same level. "Hello. Good morning," she said.

Adriano began to smile. She instructed her son to introduce himself: "I'm Tata." She laughs as she remembers his two-year-old speaking abilities.

And then, she says as though it were the most natural thing in the world, "We started to talk." From that day forward, Shalla, Tata, and Adriano became friends.

Later on, Shalla takes me to meet with Adriano without Tata, so I can get to know him myself. Like Shalla, Adriano has expressive eyes that confidently maintain contact when he speaks to you, and a charming smile that spreads across his face. He is wearing a faded purple T-shirt, black shorts, sneakers, and heavy wool socks that are out of place in the humid climate.

Adriano suggests that the three of us walk to a nearby park to speak more comfortably. I need Shalla with me, and not just because she speaks Portuguese and English. I also need her to help me understand Adriano as she does.

As we walk to the park, I notice that his sentences run

together, ideas flowing from one to another without clear beginning or end. He makes a distinct clicking noise with his mouth when he pauses. But he's an engaging speaker, challenging you to keep up and respond, even if the construction of his thoughts is not always linear. Despite the pronounced smell of sweat and the streaks of dirt across his T-shirt, he doesn't have that dishevelled homeless look. Still, Adriano stands out in the chic—and mostly white—neighbourhood.

We sit down at a green cement table, surrounded by the lushness of tropical plants and towering trees and the sounds of children playing at the nearby swing set. We begin to talk about his life.

"Ah, what I wanted in life?" Adriano muses to himself, his voice raspy and silky at once. He has a distinctive drawl. "What I wanted was to live a well-travelled life. I wanted to be someone who didn't depend on anyone, you know? To have a life that was my own, and that I understood. To survive, you know?"

His thoughts flow out of him almost like a stream of consciousness. Suddenly, he switches subjects.

"We must find a way to understand people's lives whom we've judged. You don't judge a person you just pass by and look at [because] you do not understand."

"It's true," Shalla says, nodding. She hangs on to his words, listening to and curating the twists and turns of his thoughts. Occasionally, she picks up on a thread and helps him draw out the meaning or clarify an idea.

Is he talking about making snap judgments about someone who is homeless?

"That's how the opinion of others is formed, isn't it?" Adriano continues. "So how can you give value to a person you don't know—you do not accept—if you don't know their life."

Adriano, I learn, is 28 years old and has lived on the street for the past 12 years, after leaving home due to some family difficulties that I can't get him to reveal. He is very clear—almost defensive—that this is where he is supposed to be.

"I have my life, I know what I do," he says. "I don't give any difficulty to other people, you understand? Because this is how my life is, whether people understand it or not. But I talk this way because I have capacity."

Shalla notices that I look confused by Adriano's train of thought.

She explains that Adriano uses the word *capacity* to mean an ability to think, to act, to produce. Shalla wants me to understand that Adriano has a self-awareness of his own agency.

Adriano tells me that he grew up in São Paulo, dropped out of school after completing the eighth grade, and found work as a forklift operator. He's bouncing between thoughts, but it becomes clear that his family life imposed conditions on him that he could not adjust to, and this led to a fierce determination to live by his own rules.

"My life is not accepted by everyone because I am a strange guy, I cannot . . ." his voice trails off again as though he is talking to himself. "But you don't judge a person that you don't understand. You cannot judge what you cannot do, right? I want to say that I admire a normal person, because whoever works, who has all of these things, it's not that he is better . . . but the world turns."

"That's so beautiful!" Shalla exclaims.

I think Adriano is trying to tell me he knows he is different from what he calls "normal people," and that it's okay. I ask about his family.

"I had problems in life," he says. "I was a bad guy, I ran away from where I lived, I carried a lot of drugs, I had to leave my family, I had to live in my own way."

I want to know what he thinks about the rest of us.

"What have I observed about human beings? A capacity that everyone thinks and carries with them. But I have a capacity, yes, I have the capacity to have life, I can achieve everything in life. I have capacity, I am a human being equal to all. Simply this!"

"This is beautiful, Adriano," Shalla exclaims.

"Ahhhh, thank you!" he responds, beaming, basking in her praise.

"I think people need to take advantage of what you know, of who you are, of what you have lived," Shalla adds, no longer just a listener and interpreter but a counsellor. "Your whole story, it is incredible. I admire how you live."

Adriano nods.

"Everything has the possibility to improve," Shalla urges. "You have to believe, because if you do not believe, things do not happen."

"It's true," he says quietly.

"What is your dream for the future?" I ask Adriano.

"My dream, thinking about the future?" He repeats the question, considering the words.

"I always ask him this," Shalla explains to me and Adriano, proceeding slowly. "What he wants—how we can make this happen."

"What I want for my life? I will sit alone and I will think. What I want for my life."

"It's the key," Shalla says. "For us to understand our own stories and to know what we want after."

Adriano begins talking about going to lunch and pulls out several crumpled 50 *reais* bills from his pocket. I ask him where he gets money from.

"I work. I do security down on that street."

"He guards the cars that people have parked," Shalla clarifies. "There's a beautiful story. Can I tell it, Adriano?" He nods.

One day when Shalla and Adriano were talking, he stood up suddenly and crossed the street. A man was outside of his car, frantically looking in his pockets for the key. Adriano approached the man, who didn't turn or acknowledge him. "Look," Adriano said. "Here are your keys."

As it turns out, the man had parked his car and dropped the keys by accident; Adriano had picked them up and sat waiting for the man to return. Frightened, the man thanked Adriano curtly before taking off. Shalla posted the story to Facebook, and people started approaching Adriano on the street and telling him what a wonderful thing he had done.

"You've become famous," I joke with Adriano.

"It's true," he says, smiling. "I have done this many times, guarded cars, picked up keys . . . " He trails off again, whispering.

"It's lovely to help others with their things," Shalla says patiently, pulling him back into the conversation. "It's beautiful what you do, Adriano."

This is not the first time the young woman has bonded with a homeless man. In 2011, Shalla met Raimundo Arruda Sobrinho, a homeless poet who had lived on the streets of São Paulo for 35 years. What started as a series of friendly visits evolved into a years-long relationship that changed both of their lives.

The story and its outcome are remarkable. Shalla meets Raimundo on a busy median. She finds out he's a poet and builds a Facebook page to showcase his writing. The page goes viral. Raimundo's relatives get in touch with Shalla, and soon he is reunited with his family.

Much of that story is known. What brings me to Brazil is the part of the story that hasn't been told before. The story of the young woman whose incredible gift for empathy made it all possible.

"Have you ever stopped to think that a homeless person could become one of the most important people in your life?" With this opening question, Shalla's book, *#Incondicional* (*#Unconditional*), published in 2016, takes readers on a journey through her years-long friendship with Raimundo, from their first connection to the international sensation that the story became through the press and social media. Reading the book is much like talking to Shalla in person, with her boundless admiration for Raimundo, the way she uses storytelling to recreate scenes, and the way she gently but urgently demands more from all of us. Shalla forces us to consider the possibilities for radical transformation if only we were to give more of our time to others.

Raimundo Arruda Sobrinho was born on August 1, 1938, in Goiás, a state located in the centre-west region of Brazil. In 1961, at the age of 23, Raimundo moved to São Paulo. From what Shalla can piece together about his early beginnings, he considered himself a writer. She says he intended to study at one of the city's many universities but ended up working as a gardener and a bookseller. One month in the late 1970s, he couldn't afford to pay his rent at a rooming house. By the time he and Shalla met, he had been living on the street for nearly 35 years. He'd occupied the spot where he and Shalla first met for 19 years.

In May 2011, Shalla had just moved with Ignacio to the Alto de Pinheiros neighbourhood of São Paulo, close to where she lives now in Vila Madalena. Shalla is from Niterói, a city across the bay from Rio de Janeiro; Ignacio is from Argentina. The couple met abroad and fell in love almost at first sight.

On a walk to reconnoitre their new neighbourhood, Shalla and Ignacio found themselves passing along the Avenida Pedroso de Morais, a grand boulevard dotted with well-to-do houses that ends at the University of São Paulo, its two-way traffic separated by a generous median with trees, a running trail, and a path for cyclists.

Shalla didn't know it at the time, but Raimundo had named the median where he lived for 19 years "the Island." Sitting in the middle of the median was a tent made entirely of small plastic bags used to hold rice that had been stitched together into what Shalla calls a cabin. And beside the cabin, seated on a couple of paint cans, sat a person wearing a garment made of the same plastic bags.

Shalla takes me to the exact spot where she first set eyes upon Raimundo, where the grass now meets a concrete pad. Homeless people often seem to just sit. To Shalla, Raimundo appeared to be doing something purposeful, even noble.

"He was concentrated in what he was doing," she recalls. "As you can tell, it's a noisy place with a lot of cars and traffic and buzz. I did not know if it was a woman or a man because the plastic bags covered the face that was framed in long hair that was matted and stringy. And yet, he was writing and writing."

Shalla stopped walking and stared. "Wait," she told Ignacio. "That person is enlightened. I want to talk to him. Or her."

"Shalla, maybe this person doesn't want to talk to you," Ignacio told her gently. "People who have lived a long time in the street are in their own world, physically and mentally. Maybe it's not a nice idea."

Shalla says she and Ignacio were in a rush, and so she dropped the idea of stopping, but for the next two days she thought of little else. On the third day, Shalla decided she had to try to make contact. She returned to the spot where the cabin was still standing.

"I crossed the *avenida* and stood there looking," Shalla recalls. "I put my feet on the person's land. I approached very slowly because I knew I was getting into this person's house. Then I stopped right here [smack in the middle of the median] and I waited to see if I was welcome."

Shalla greeted the person with "good morning," and her salutation was met with a smile.

"Do you want a piece of paper?" the person asked Shalla.

"Sure," Shalla told him and stood waiting.

She was handed a scrap of paper. There were words hand-written on it.

"May I know your name?" Shalla asked, accepting the paper.

"I am Raimundo Arruda Sobrinho," he said in a formal tone of voice.

He's definitely a man, Shalla concluded.

"What is your name?" the man asked.

"My name is Shalla Monteiro," she replied. "I saw you before. I didn't know if I was welcome to come, but I really wanted to know you and tell you that I was thinking about you."

Shalla says that Raimundo continued to look at her, enjoying the moment. She recalls looking at the piece of paper and realizing it was a poem.

"Raimundo, can I read in a loud voice what you have written—to make sure I understand what you are saying here?" Shalla asked him in a respectful tone of voice.

Raimundo continued to stare at Shalla. She sat down on the ground beside him and started to read the poem aloud: *"O que é o leitor interessado sobre a vida exterior e os outros consumidores sobre toda a quem humana tem feito."*

Shalla translates the words for me:

"What is the interest of the reader,

For the life of the author he has read?"

To Shalla, the words suggested that Raimundo had empathy for a carefully selected reader of the poem. For her.

"Do you believe he just happened to be writing that poem when you came along?" I ask.

"I cannot tell you if he was finishing that poem exactly at the point he gave it to me or if he pulled it from a stack of poems he had written. For me, when I read it, I knew that I must know him," she says. "The energy, the connection, whatever we call it. I felt it from inside my body."

That day, she stayed with Raimundo for a long time. Maybe an hour, maybe more. She says she lost track of time. He told Shalla his life story. She told him hers. A bond had been created. "For me, it was like he's a master," Shalla tells me. "I wanted to know this man. I wanted to come here every day."

And so she did. Sometimes more than once a day; sometimes even late at night. If it rained, she would get in the car and drive by the Island to see if Raimundo was getting wet and bring him towels to dry off.

They spoke often. She says their conversations were deep and philosophical. She was surprised at how well-read Raimundo was. He told her he'd read 3,000 books. Still, he had his peculiar thoughts. For one thing, he told Shalla he had a radio transmitter in his head. For another, he described himself to her not as a human being but as an animal.

"What do you think that means?" I ask.

"He told me he's an animal because he is accustomed to living in worse conditions than you or I," she says.

When she asked about his family, he would change the subject.

"I'm a rational animal," Shalla recalls Raimundo telling her. "I cannot live in a family with kids in the same house. Look at me. I cannot."

The day they first met, Raimundo told Shalla his birthday was the first of August. She decided to make a party for him. Shalla found out his favourite food and brought it. She brought balloons. Seeing that he had difficulty getting around, she bought him a walker. But she says the most important gift she gave him was a poem. She used the same kind of paper Raimundo used and wrote it out with the same kind of pen. She copied meticulously his writing style.

"I wrote about how important it is for me to read his words and to get to know him," Shalla recalls. "At the end I wrote, 'I love you.' He cried. Everyone cried. Ignacio got emotional too. That day, we had the first physical contact. It was a strong hug and a kiss on the forehead."

The emotional connection was very important to their relationship. That kiss on the forehead told Raimundo he was more than the stereotype of a homeless man. He was a person worth getting to know. And love.

After spending the morning with Adriano, Shalla and I drive to Hospital das Clinicas, the sprawling medical complex of the University of São Paulo. Our destination is the Institute of Psychiatry, an imposing building renovated in the early 2000s, with a glass entryway that juts diagonally and dramatically downwards. The institute is among the largest and most prestigious university hospitals in psychiatry in Latin America.

Shalla and I meet with Dr. Valentim Gentil, an esteemed psychiatrist who holds a medical degree from the University of

São Paulo and a PhD from the Institute of Psychiatry at King's College in London. He has been a professor of psychiatry at the medical school for decades and is renowned in the field.

Dr. Gentil also has a personal connection to Shalla's story: He has known Raimundo since the 1990s. When a friend showed him the Facebook page Shalla made for Raimundo, Dr. Gentil reached out to her to meet and became the first person to encourage Shalla to write her book.

Dr. Gentil tells me how he came to meet Raimundo.

"I lived in the area where Raimundo had his base and I would pass by him at least three, four times a week," Dr. Gentil says. "On one of these encounters, I asked to take a picture of him and to talk with him."

As with Shalla, Raimundo handed the doctor a "philosophically minded" poem. The doctor told Raimundo that if he ever wanted to return to the Hospital das Clinicas, where he had been briefly admitted in 1976, he would be pleased to look after him.

"Oh, thank you very much, Doctor Valentim," Raimundo responded.

"Wow!" says Gentil now, laughing. "Half an hour later, he still remembered my first name. I thought that this man doesn't have major cognitive disturbances."

Many of São Paulo's homeless population live at least some of the time at one of the city's 79 overnight shelters. There are currently about 10,000 vacancies for an estimated homeless population of 16,000. According to the latest government census, some 7,000 people, like Raimundo, live entirely on the street.

"What is worrisome—and this corresponds to Raimundo's problem—is that a number of these people have serious mental illnesses," Dr. Gentil says. Data is difficult to come by, but based on available statistics, he believes that more than 10 percent of the homeless population has a severe mental illness and has gone without treatment for at least one year.

"Many people with major psychiatric disorders, problematic drug use, and alcoholism remain in the streets overnight, both in the summer and the winter," he adds. That can lead to major health problems like tuberculosis and heart disease.

"I used to call ambulances. The ambulance would arrive, the police would come, and they would take the patient to an emergency room. But the emergency department wouldn't have anywhere to refer the patients to."

The problem, Dr. Gentil says, is that in Brazil, more than 80 percent of beds in psychiatric hospitals were closed over 20 years in what he considers a poorly thought-out move away from institutionalization. The situation is true of Canada and other industrialized nations. "I think the idea that we would be able to provide psychosocial rehabilitation and deinstitutionalize severely mental patients was a fantasy," says the psychiatrist.

But no matter the challenges faced by the population at large, Raimundo managed to build something of a steady life for himself on the street, a remarkable feat, Dr. Gentil says. "He has a chronic delusional system, but it was amazing that he was not disturbed by the other people living in the streets or by the population around. The police did not interfere with his living on that small island for so many years.

"He was a character. He belonged to the neighbourhood; he belonged to the community there," Dr. Gentil adds. "Everyone knew him, perhaps not by name, but everyone knew who he was."

I ask Dr. Gentil about Raimundo's assertion that he had a "radio" installed in his head. "Yes, he told me that when I met him a number of years ago—he told me that he came to this hospital and we implanted a device in his brain."

Raimundo told the doctor that he believed that his "thoughts were stolen" and that information was being transmitted to him through this device.

"[This] is a disorder of perception. It's a thing described in psychotic patients of the schizophrenia or paranoid types," Dr. Gentil says. "We don't really know how they feel it, but it's kind of a lack of a barrier between what they think and what they think they hear."

Still, compared to many with schizophrenia, Raimundo was easier to talk to, given his ability to communicate, both verbally and through his writing. Often people with hallucinations or delusions mistake the intentions of others, leading them to become fearful.

"I think, in the case of Raimundo, some people in the neighbourhood where he lived knew him or would acknowledge his presence and would try to help, but the society as a whole didn't provide what was really needed," Dr. Gentil says.

To illustrate his point, he tells a story of leaving the supermarket and seeing an older person who had fallen and was bleeding. A swarm of concerned people had gathered around him in the main entrance, calling an ambulance and offering help. As Dr. Gentil walked farther away from the supermarket, he noticed another person in distress just a metre or so away. This man was likely intoxicated and unconscious. He was also homeless. "The man was at risk of all sorts of medical problems, but there was nobody around. Nobody would call an ambulance for him."

The disparity was troubling to Dr. Gentil. He even wrote to Brazil's medical board, the Conselho Federal de Medicina, and asked if there was an ethical obligation to offer help. He says the board replied that the problem is so widespread that he "cannot be personally responsible for it."

"The attitude of people here and elsewhere toward severe mental disorders is to fear them, to stigmatize them, or to have complete ignorance of the illness and how to respond," he says. "And health services don't do their part."

But Shalla did; first, by ignoring—celebrating may be a better word—Raimundo's quirks and peculiarities that frightened others over the years, and second, by creating a bridge not just between Raimundo and herself but also between the writer and his audience.

As Shalla got to know Raimundo better, she learned that he had no interest in receiving charity from her.

"When I would eat something in my house, I would put the leftovers in a Tupperware and bring it to him. Or little presents. One day I said, 'Please, accept this,' a fifty *reais* bill. On that day he stood up and walked over to me. He said, 'Please, I really don't want it.'"

Shalla says that Raimundo survived for decades on the streets of São Paulo by doing odd jobs and by being a trader. But he always thought of himself as a poet. The irony is that the part of town where Raimundo built his cabin made of plastic bags is where many of the city's most successful editors and publishers live.

"Raimundo tried to show his poems but his payment for doing a chore was a plate of food," Shalla recalls. "He would say, 'I don't want food. I want to show you my poems.'"

Raimundo's poems were his currency, and a much devalued one at that. Shalla recalls a well-to-do man who would show up with his car and driver to drop off some food. "I saw him more than 10 times," Shalla recalls. "To give Raimundo food was his good deed." In return, Raimundo would always give the man a poem.

"The man would never physically touch Raimundo," she recalls. "One day, I went to his car because he had come without the driver. He said he didn't want to touch the paper with the poem because it had been touched by Raimundo. For hygienic reasons." Shalla shakes her head.

Shalla asked other people who had interacted with Raimundo if they were aware that the scraps of paper he handed them contained poetry. Most said they weren't.

It is at this part of the story that Shalla's empathic qualities and her occupation intersected. Shalla has a degree in advertising; her husband, Ignacio, has a degree in social anthropology. Tree Intelligence, the company they jointly founded in 2011, uses data analytics and a proprietary knowledge discovery platform to help clients identify everything from consumer trends to the influence of digital media.

Shalla decided to bypass the revulsion that passersby felt about Raimundo by giving him a persona that would entice people instead. She put Raimundo and a selection of his most vivid and powerful poems on Facebook. Before she did it, she brought Raimundo up to speed.

"I explained everything about the Internet," she tells me. "He is very smart. He got it right away. Together, we chose his profile picture—a nice one of Raimundo taken at sunset. Then, we gathered a portfolio of his poems. We selected some of them, and then I put together a photo album along with a few of Raimundo's words in large type, plus the photos of his cabin."

Shalla launched the page. It went viral. Today, the site has close to 200,000 likes and nearly 200,000 followers. It reframed the way people thought of Raimundo. Strangers wanted to stop by to meet the homeless man with a page on Facebook.

"Hey, Raimundo, I lived here for eight years," Shalla recalls one person saying. "I always thought about you, but I never had the courage to talk to you."

The page and the poet were profiled in newspapers and on television. Then, something even more amazing happened: Shalla received a message from one of Raimundo's brothers, Francisco Arruda Sobrinho. "Raimundo," he wrote, "I am your brother. We are looking for you, we want you close to us."

"I couldn't believe that we had found his family," recalls Shalla, shaking her head. "Imagine that you lost contact with your family for 51 years, and then someone comes and says, 'Brian, I found your brothers and sisters.'"

After verifying the brother's story, Shalla sat down with Raimundo to give him the news. Raimundo's reaction was typical of him.

Are they happy? she recalls him asking her. "That was the only thing he asked. He is so amazing."

Over the ensuing weeks and months, Shalla became the point of contact between the brother and Raimundo. Francisco and his wife travelled to São Paulo to visit with Raimundo, beginning a long process of reconciliation and eventually of reuniting Raimundo with his family.

Together, Shalla and Raimundo's family worked with São Paulo's department of social assistance to develop a plan for reintegrating Raimundo into the "normal" world after 35 years of living on the street. On April 23, 2012, Shalla and a team from the city brought Raimundo to a psychosocial care centre, where he would live temporarily while he adjusted to life off the streets. He had to get used to everything, from taking a shower (he had only taken two showers in decades) to wearing clean clothes and interacting with others.

In July 2013, after 51 years away, Raimundo returned to the

state where he was born, Goiás, to live with his brother Francisco and his family.

To reach Raimundo, Shalla had to cross a busy boulevard to build a bridge to the island where he lived. Instinctively she understood that to get along with Raimundo, she had to flip perspectives and see things from the poet's point of view.

She also knew how to use Facebook to build a bridge between the poet who lived in a cabin made of garbage bags and his family, not to mention thousands of readers. But what gave her the instinct to build that bridge and to cross it in the first place? Dr. Valentim Gentil, the psychiatrist who knows both Shalla and Raimundo, has an interesting perspective.

He and Shalla met after a colleague found Raimundo's Facebook page. "I don't have Facebook, but I thought, 'Oh, this is fantastic, I must meet her!'" Dr. Gentil says.

They did meet, and Dr. Gentil says she didn't disappoint. "It's amazing, we have to clone her." He laughs. "It's more difficult to understand what happened with Shalla than what happened with Raimundo. I wish that we had more like her."

So what does he think sets Shalla apart?

"She's obviously a courageous person," he says. "She's sensitive. She has this feeling of fraternity. She's very humane."

But, he says, it's more than that.

"She has the ability to transform ideas into actions very effectively," he adds, noting that there were a number of situations during the initial process of moving Raimundo off the street when Shalla "had to use all her diplomatic and political ability to get all those people who helped her to do their job."

He too believes her background in communications and

publicity means she knows how to express things in ways that get people to pay attention. "The idea of the Facebook page was fantastic! And her level of commitment, more than most people would have given. But I still don't understand," he continues, laughing again. "Perhaps she could provide a training for society."

I joke that we should study Shalla. "Not for study," he corrects me. "She should come and teach us."

I am still curious about what in Shalla's background made her comfortable enough to approach a homeless person with a mental illness like Raimundo and develop a deep relationship.

"You better ask her," he responds, laughing.

Shalla takes me on a special walking tour of São Paulo. We're not visiting museums and other tourist attractions. She's taking me to places of significance during Raimundo's 35-plus years in the city.

"This is the first place that Raimundo lived when he arrived in São Paulo," says Shalla, pointing to what is now a supermarket. Back then, it was a rooming house for itinerant men. "He was 21 years old. He came here with the dream to study. He had a friend that came before him and said, 'Okay. Let's come to São Paulo. There you can study.'"

Shalla looks dreamily at the supermarket, willing herself to picture the rooming house back then and Raimundo as a young man with hopes and dreams and a life yet to be lived.

I ask Shalla if she thinks of herself as an empathic woman.

"I don't believe, how do you say it?, that I have empathy," she says. "But maybe I have this perception or feeling of others. People tell me things that are very deep inside them."

If you have ever met a total stranger and found yourself telling that person your life story, it may have been someone like Shalla.

Once, while travelling with her business associates to attend a conference in Milan, Shalla stopped at Piazza del Duomo, the square in the centre of the city dominated by the Milan Cathedral. Shalla was sitting alone, contemplating the square, when a woman saw her and sat down beside her. Shalla was just 28 at the time. The woman was in her fifties. They started to talk.

"She told me a lot of things about her life," Shalla recalls. "She told me that she was married to the wrong guy and that she was unhappy. She said that maybe she transferred her unhappiness onto her daughter, and that perhaps her relationship with her daughter was not good because of that. In deciding to get married, she abandoned an opportunity to go to another place."

The woman confided in her for two solid hours. "She told me she never had this deep conversation with anyone else," says Shalla. "She thanked me for being her psychologist."

Shalla is quick to point out that she is not a psychologist. Still, she was so engrossed in listening to the woman that she forgot the meeting she was supposed to attend with her coworkers.

"You were in the moment with her," I tell Shalla.

"I was there," says Shalla. "Completely there."

She was also "completely there" when she met Raimundo.

"When I saw Raimundo for the first time, I couldn't see anything other than [the fact that] he was enlightened," she says. "He was meditating. It was something that I cannot explain, because it's not rational. It's like when I met my husband, Ignacio. We were in Mexico City, in a very huge square called the Zócalo."

Today, the Zócalo is also known by its formal name, Plaza de la Constitución, or Constitution Square. Before the modern state, it was an Aztec ceremonial site.

"It's a very symbolic place, and I was there," Shalla recalls. "A lot of people are there. I saw Ignacio Garcia, and he saw me, and we were connected. That day, when we saw each other, he came to talk to me. I don't know how I knew, but I felt that he was going to be the guy who I would marry."

Shalla and Ignacio married soon after. As with Raimundo, Shalla established an instant connection with Ignacio. Interes-tingly, she calls Ignacio her husband and Raimundo her soulmate. She can't explain the difference; she just knows it on an intuitive level.

Shalla takes me through a forested area of São Paulo. "We are very close to the Island," says Shalla. "Very close to the house I used to live in when I met Raimundo."

Living in the moment can be intoxicating. When I visited Santiago de Chile during the *estado de sitio*, the state of siege declared by Chilean dictator Augusto Pinochet in 1985, I fell head over heels for a beautiful young woman I met at the offices of a human rights watchdog group. So afraid was I of losing control that I quickly broke it off.

For much of my adult life, I have arranged things so I know exactly what will happen next. I find it very difficult to let myself just be in the moment; for Shalla, the moment is exactly where she wants to be. My intuition says the difference is important to discovering where Shalla's ability to connect comes from, and where mine might have gone.

Shalla was born in Niterói, a small city that faces Rio de Janeiro across Guanabara Bay. The oldest of three siblings, she says her life as a child was "very normal." Then, in the next breath, she says that when she was 13 years old, her parents got divorced.

I ask how that affected her.

"Actually, I loved it," she says with a perplexing smile. "They were fighting all the time since I was five. They were not *connected*, you know?"

Shalla says that from age five until her parents got divorced, she acted as their go-between. That may have forced her to grow up before she wanted to.

That happened with me, albeit for different reasons. From an early age, I felt the weight of responsibility to do something with my life in order to make my mother proud.

There was also a recurrent pattern during Shalla's childhood that began when she was six or seven years old. "We'd get ready for a trip," she says. "We put all the things in the car to travel. Then, something would happen. Something little, like we'd be stuck in traffic. My parents would start fighting, and my father would drive us back to the house. The trip was over."

Parties. Days at the beach. Happy anticipation followed by crushing disappointment. She learned never to look forward to pleasant times that could be snatched away suddenly and capriciously. "I learned that I'm not expecting a lot of things from life," she says. "If good things happen, whoo! But if not, that's okay. I will do what I can."

But that's the grown-up Shalla looking back on her 13-year-old self. As a kid, those sudden reversals, coupled with her parents' acrimony, may have made her childhood bitter. But she found a way to shut out the pain. Shalla discovered that she could enter a state of meditative bliss in which she could push the trauma of her childhood away.

"I can be in very deep contact with myself," she says. "I can be quiet, without thoughts. When I go to sleep, I can say to myself, 'Here, it's nice inside.'"

Later on, Shalla went to yoga classes and had formal training in how to meditate. But there was nothing in those practices that she didn't already know.

Being able to enter that state of what she calls *deep contact* almost at will has given her an ability to cope with stress that is far better than that of her husband, Ignacio. It has also given Shalla an uncommon outlook on life. If there are only regrets from the past, and nothing to look forward to, then what else remains?

"You can only depend on the moment and how you react to the moment," says Shalla. "You can react in a lot of ways. You can accept it. You can fight. You can be disgusted. It all depends on the context."

When her parents separated, Shalla did something extra-ordinary. She gathered her younger brother and sister up in their beautiful apartment in Niterói, with a spectacular view across the bay looking at Rio, and said to them, "Let's go look at the view for the last time."

As a 13-year-old, Shalla had learned to live in the moment.

Now, I understand why she reached out to Raimundo. She saw someone ignoring the pain and degradation of homelessness to strike a pose of enlightened meditation. She saw a soul who lived in the moment. She saw herself.

Today, Raimundo is settled into his new life with his brother and his extended family. He and Shalla speak to one another via Skype.

Inspired by Raimundo's Facebook page, several projects have emerged on social media platforms in Brazil to change perspectives on homelessness. "São Paulo Invisível" (Invisible São Paulo) is a Facebook initiative started in March 2014. The project's goal is to "open eyes and minds through the stories of the invisible to inspire people to have a more human view," drawing people closer to those lives that are so often neglected, ignored, feared, and remote in São Paulo.

"What I want to do is to encourage people to develop emotional connections to those who are homeless, to develop friend-ships," says Shalla. "I think that is my path, you know?"

Dr. Valentim Gentil urges Shalla to help others. "You have to use this energy," he tells her.

She does use that energy to help other homeless people. But they're not Raimundo. They never will be.

"Let's go to the park?" Shalla asks her son. "*Vamos! Vamos! Vamos!*" Tata sings. His dark eyes sparkle and dimples punctu-ate his round face.

Adriano, the homeless man who has bonded to Tata, decides to remain on his self-declared territory on the street corner. Shalla, Tata, and I say goodbye and begin making our way to the park.

When we reach the park, Tata stops at a towering tree, staring up at it and feeling the bark textures beneath his little hands. It reminds me of Shalla's desire to live in every moment.

So where does Tata get his empathy from? I ask Shalla.

"I think it can be inherited, though I don't know if it's neces-sarily genetic, but rather through the experience of living with someone," she says, pushing her son on a set of swings. "The form in which the child perceives the other. Children absorb a lot from their environment. Certain traits are their own being, of course. But I think that there is something about their environ-ment that either stimulates or represses certain things."

Perhaps Shalla has passed something of her eagerness and openness to the world, to new people and situations, on to Tata. "This question is looking at another and not judging, but rather, trying to connect. Perhaps this is a root of empathy."

So how can we teach this quality in families, in schools, in our society? Aside from obvious issues of safety, Shalla has a very open style of parenting. She doesn't tell Tata what he can and cannot do. She just lets him explore the world. In their apartment, there is no television, and among Tata's various projects is a cooking station with real vegetables and little plastic knives for cutting.

"It's like this little park we are in," Shalla explains, which has an enclosed fence and childproof gates. "In here, the children are free, but when you leave this area, the child can no longer be open. I hope Tata doesn't have that sense."

At this phase of beginning to set limits, Shalla tries to give Tata the confidence to continue exploring, to improvise, to move naturally through the world. "My philosophy is to preserve the *being*," she tells me.

She says there is some discussion about whether Tata understands that Adriano is homeless, but mostly it doesn't seem to matter. We sit in a sand pit while Tata plays with trucks, and I ask him again about his friendship with Adriano.

"I play with him. I talk with him." Tata runs his trucks through the sand, accompanied by sound effects.

"How did you become friends?" I ask.

"I gave him a kiss and we became friends," he replies without looking up. "This truck removes the sand. Look, Mom!" he starts shrieking.

Is Adriano the same or different from you?

"The same."

And why is that?

Tata pauses to think while flying one truck over the other.

"Because he's beautiful."

CHAPTER EIGHT

The Kindest Robots

My personal search for empathy has taken me to meet people who are born kind, people who have acquired kindness by facing pain and disappointment in their lives, and people who have had compassion thrust upon them. In a world bursting with technology, there's another kind of compassion worth exploring. Call it kindness that is manufactured from pneumatically actuated joints, smooth silicon skin, printed circuits, and heuristic algorithms. It's no wonder that my search for kind robots has taken me to Japan. The country is crazy about robots and androids. They're everywhere.

"Pepper" is a humanoid robot companion developed by Aldebaran Robotics that is designed to communicate with people and to read emotions ranging from joy to sadness. Words, tone of voice, facial expressions, and even the angle of your head can tell Pepper what kind of day you're having. For $1,600 plus $360 a month on a 36-month contract, you can take Pepper home with you, something 1,000 Japanese customers a month are doing.

In 2015, Toyota introduced the Kirobo Mini, a pint-sized robot that fits in the palm of your hand. It vocalizes like a baby, blinks its huge eyes, and recognizes facial expressions. Like an infant, it can sit up but flops over easily.

At Miraikan, the National Museum of Emerging Science and Innovation that overlooks Tokyo Bay, Japanese school kids clap deliriously as ASIMO goes through his act. ASIMO (Advanced Step in Innovative MObility) is a humanoid robot built by Honda. Every day at 2 P.M., the robot jogs, dances, waves his arms, and sings in Japanese and English.

If anyone has learned enough about empathy's secrets to imprint them onto machines, it's the Japanese.

"Konnichiwa, Hasegawa-san." Kiyoko Kitagawa calls out a loud hello to Mr. Hasegawa as she raps on the front door. The case manager leans forward and listens intently. *"Konnichiwa,"* a thin, quiet voice replies from inside the building.

Kitagawa is taking me to meet Tomio Hasegawa at his ground-floor apartment in the Nishinari Ward of Osaka City. It's a poor neighbourhood best known for housing Osaka's day labourers, many of whom work in construction. That's what Hasegawa did until a back injury forced him into retirement from his job as an assistant carpenter. Today, he is one of an estimated 25,000 elderly retired day labourers living in Osaka. The case manager says that Mr. Hasegawa is fortunate to have his own home; an estimated 1,300 of Nishinari's elderly retired workers are homeless.

Kitagawa unlocks the front door and slides it to the right. We enter a small vestibule, where Kitagawa motions for me to take my shoes off. In Japan, it's a sign of respect and a necessity to keep floors clean in homes where people eat on very low tables. We enter a larger room to the left of the vestibule.

"*Ohayō gozaimasu.*" Mr. Hasegawa wishes us good morning in a slurred voice thick with phlegm.

Mr. Hasegawa is 76 years old but looks a lot older. I find him bed-bound on a hospital bed that dominates the room. He seldom leaves the house.

"I have 34 clients on my caseload," says Kitagawa. "Mr. Hasegawa is by far the most needy client I have."

The living space is crammed with nursing home equipment. To the right of Hasegawa's hospital bed is a tray table on wheels, a commode, and a straight-back chair. On the wall to the far right is a three-drawer clear-plastic cabinet filled with gauze pads and other supplies.

"It's a little cramped," Hasegawa says in Japanese, his mouth forming a toothless smile. It's quite the understatement. The apartment where he's lived since his stroke is all of 132 square feet. The only clues of life beyond his medical needs are a calendar filled with photos of sumo wrestlers and a small flat-screen TV for watching sumo wrestling and baseball.

"Pretty soon, the baseball season is going to start," he says.

"Have you heard of the Toronto Blue Jays?" I ask.

"The Hanshin Tigers are better," he says.

The Hanshin Tigers, based in Osaka, are one of the oldest professional ball clubs in Japan. The team is named after the Hanshin Electric Railway Company, which bought the team in 1961.

"Cecil Fielder was a good ball player," says Hasegawa.

I knew we'd find something in common. Cecil Fielder played for the Toronto Blue Jays from 1985 to 1988. In 1989, he left the Jays to play for the Hanshin Tigers before returning to North America.

Hanshin Tigers fans are known for standing up en masse and making noise with kazoos while waving the team flag. During the seventh inning stretch, they release hundreds of balloons while singing the Tigers' fight song. When a player comes up to bat, a brass band plays a song written especially for the player, and the fans sing along.

"Sometimes, I sing along while watching the game on TV," says Hasegawa.

I move a bit closer to the old man. His skin is pearly grey and thin. A stroke has left him partly paralyzed. His elbow is flexed, and his gnarled right hand is bent into a fist so tight that his fingers dig into the palm of his hand. Arthritis in his spine has left him in constant pain. He is unable to walk. He can't chew solid food, and so he lives on a diet of pureed rice, fruits, and vegetables. He can't get out of bed without assistance. Like a growing number of frail elderly people in Japan, Mr. Hasegawa needs a lot of help to stay at home. It's Kitagawa's job to line up supports to keep him there.

"Even with his health deteriorating over time, he would still try to live independently by himself," she says. "If that is his will, then we want to assist him with that lifestyle. And if one day, he decides it's time to go to a nursing home, then we will assist him to do so."

A personal attendant visits Mr. Hasegawa three times a day to clean and wash him, feed him, give him his prescription medications, and get him in and out of his chair.

Like most retired day labourers, Mr. Hasegawa is almost penniless. After paying for food and a few personal items, he has nothing left over. Kitagawa says the hospital bed is paid for by Kokumin-Kenkō-Hoken, the national health insurance plan for people who aren't eligible for employment-based health insurance.

I ask Mr. Hasegawa if he has family to look after him. "I am the youngest of eight brothers and sisters," he tells me. "I'm the last to go."

Mr. Hasegawa says he never married and has no children to look after him. He has never met his nieces and nephews. Most retired day labourers in Osaka have no connection to their family of origin. They live alone and die alone, with no next of kin to claim the body for burial.

Kitagawa says the big challenge is getting the government to pay for hired help. "The government expects people to provide for themselves," she says. "Often, his personal attendant works for two hours but only charges for one."

"Why would he do that?" I ask.

"He likes him," says Kitagawa. "He doesn't like to see him suffer."

Japan is sitting on a demographic time bomb. Already, it has the oldest population on the planet. More than one in four people living here are age 65 and up. People in Japan live longer than citizens in most other developed nations. Even retired day labourers like Mr. Hasegawa can expect to live to age 80 and beyond. A low birth rate in Japan means fewer working-age people and less tax revenue to care for seniors.

By 2050, four out of every 10 people here will be eligible to receive a pension. And, like Mr. Hasegawa, many of them will need a lot of help. By 2040, more than one in four seniors in Japan will have three or more limitations with things like shopping and getting to doctors' appointments. Like Mr. Hasegawa, a little under one in four will have trouble doing basic things like getting dressed, bathing, and feeding themselves. One in five seniors will have anything from noticeable forgetfulness to obvious dementia.

To Canada and other developed nations, Japan is the model

for how to cope with an aging population. But insiders tell a different story.

"I think Japan should have been prepared for the coming aging society," says Michihiko Tokoro, professor of social policy at Osaka City University. "But the pace of aging of the population is much faster than the government or society expected."

Nursing homes in Japan have strict quality standards, but the government can't build them fast enough. As a result, many frail seniors move into rooming houses and dormitories that are hastily converted into nursing homes. "Some of them are quite bad," says Tokoro. "They are attracting a lot of older people and getting a lot of money from families and from the social security system."

Many live at home with no family members to look after them. Smaller families mean fewer kids to look after their parents. Since most people work full time, they have little if any time to care for aging parents.

That trend has been evident for decades. The problem, says Tokoro, is that successive governments haven't kept up with social trends. To a staggering extent, the government acts like it's 1960 and expects family members to care for aging relatives.

Tokoro says that, roughly speaking, there are two kinds of seniors: those with the financial means to hire help and those without. "Governments provide care in the public system," he says. "The government assumes that families will provide 20 to 40 percent of the care. The problem is, the number of people who need 100 percent care is increasing."

Tokoro says the system needs to be restructured to meet the growing need.

"The simple answer is that we need more money from higher taxation," he says. "But that is difficult for policy-makers because it's very hard to get elected on a promise of higher taxes. If kept at present levels, the Japanese social care system cannot be sustained."

And if that happens?

"For lower-income people like Mr. Hasegawa, higher-care needs will be given priority," he says. "Middle-class and rich seniors will have to depend on themselves without relying on the public social care system." That means higher user fees for middle-class seniors.

Tokoro says higher taxes would be a start, but there's another major challenge that Japan needs to face. There aren't nearly enough care providers like the ones who visit Mr. Hasegawa.

In 2013, Japan's Health, Labor and Welfare ministry estimated that there were 1.71 million nursing care workers in Japan. By 2025, 2.53 million workers will be needed. A report by Merrill Lynch predicts Japan will eventually be a million workers short. That comes as no surprise to the Osaka City University professor.

"Their working conditions are terrible," says Tokoro. "Most earn low wages. Even if you made the wages better, the working hours are quite difficult. They have to work nights. Clients with dementia require special attention. The job might be okay for care workers in their twenties, but not when they get married and have children of their own. They end up sacrificing their own families for their jobs. They should be admired. I think Japanese society should realize how much personal attendants do to sustain the social care system."

Tokoro says Japanese culture prizes self-sacrifice, and that may be holding back personal attendants from demanding better wages and benefits. "Japanese care workers don't complain about their wages because they think this is not just a job but their social duty."

With the pressure of an aging population, governments of

several countries are embracing the idea of recruiting workers from abroad. In Canada, we recruit nurses and other health care providers from places like the Philippines to care for ailing seniors. They come to earn money and to gain citizenship so they can bring their families. Tokoro says Japan is just beginning to explore the option of recruiting care workers from the Philippines. But Japanese authorities want guest workers, not future citizens. Three years in, and foreign workers are sent packing. The second and more important barrier is cultural. Japanese people tend to be insular.

"We need more workers, but we need workers who speak Japanese and who understand our culture," he says. "The problem is not with the capabilities of workers from abroad. The problem is with Japanese society itself. I'm not confident that our society is ready to invite these people."

Rather than solve these kinds of problems, Tokoro says the government is trying to encourage more people in Japan to volunteer to care for older relatives and friends. But he doubts that will work. "I don't think it's possible to wind back the calendar to the 1960s," he says.

Government social policy toward seniors may be rooted in the past. At the same time, Japan has invested heavily in a race to fill the gap in human attendants with robots made of silicone and circuit boards.

Like many robots on public display at museums and trade shows in Japan, ASIMO, built by Honda, is little more than a singing and dancing novelty act. But the car company wants to tap into the potentially lucrative seniors market for machines that are dubbed carebots.

Honda and other Japanese companies are trying to lead a global market for personal robots that could be worth as much as $17.4 billion US by 2020. Some of these robots amuse seniors and keep them company. These are known as social robots. Others will help people like Mr. Hasegawa perform his daily activities.

RIBA, short for Robot for Interactive Body Assistance, is a nursing care robot developed by the government-funded company RIKEN. RIBA can transfer a senior from bed to wheelchair and back. It has strong human-like arms and a sophisticated guidance system. Resyone, developed by Panasonic, is a carebot that converts from a bed into an electronic wheelchair. Either of these devices could help Mr. Hasegawa, the elderly man I met in Osaka City.

At a lab at Waseda University in Tokyo, Shigeki Sugano, a mechanical engineer and one of Japan's leading robotics gurus, has spent close to 20 years developing a prototype that could do even more. Its name is TWENDY-ONE, and it's built to assist older people at home or in long-term care. Sugano calls it a "human-symbiotic robot" because it's designed to co-exist with its owner. Sugano is the project leader, but he's had help from a large team of researchers along with the support of 20 private corporations.

Like RIBA and Resyone, TWENDY-ONE can transfer a senior from bed to wheelchair and back. But it does much more than that. "TWENDY-ONE can prepare breakfast," he says. "It can open the refrigerator and take something out. It can toast bread, and then use tongs to pick up the toast and put it on a plate."

The robot doesn't look like a human with its high-tech white finish and hands that are bright red—but it is humanoid. It has a wide, triangular-shaped head that makes it look like the title character in the Steven Spielberg film *E.T. the Extra-Terrestrial*. The eyes are CCD cameras surrounded by LEDs that light up when TWENDY-ONE's vision system is activated. The robot has a small torso that twists in several directions. It has enough power to lift a person out of bed. Two oversized arms have moveable joints at the shoulders and elbows. The hands have four fingers with lots of moveable joints. Each fingertip has sensors that enable the robot to detect and pick up objects as varied as a pencil or a coin. It can hold a disposable cup without crushing it.

TWENDY-ONE is a stupendous engineering achievement, but it won't be coming to a nursing home, let alone an elderly person's residence, anytime soon. The estimated production cost of a single robot is as much as 20 million yen or $236,000 Canadian.

Sugano likens TWENDY-ONE to concept cars that draw crowds at auto shows. "It will probably take 20 or 30 years until we have a commercial version." He says there's another problem with TWENDY-ONE. Unlike the personal attendant it's supposed to replace, this machine can only do what it's programmed to do. And it's built to assist a senior of average build.

Mr. Hasegawa is malnourished and underweight, so TWENDY-ONE would have little difficulty lifting him out of bed. However, it wouldn't be able to lift a client who is morbidly obese. It may not be able to lift a client who is unusually short or unusually tall.

That's just height and weight. Don't forget that seniors often have pre-existing injuries and disabilities. The client with a torn shoulder muscle might not be able to raise her arms so that TWENDY-ONE can lift her. Arthritis, strokes, prior surgery, heart failure, malnutrition, quadriplegia—these and other factors will make each client's interaction with a robot unique. And let's not forget that frail seniors are likely to become more disabled over time. The robot who helps a senior today may be unhelpful or even dangerous to that senior in the future.

So, Sugano is working on a prototype of a new robot that leaves TWENDY-ONE in the dust. It's so new it doesn't even have a name, and yet it puts Sugano a big step closer to achieving true symbiosis between robot and human.

Sugano's new computerized robot can adapt to the needs of a particular senior thanks to something computer engineers call *deep learning.* Humans learn and improve through trial and error. We do that by accessing vast numbers of networked cells inside our brain. These networks enable us to refine what we do by recognizing patterns in response to our actions. Deep learning (sometimes called hierarchical or deep machine learning) does something similar. It uses networks of computing power arranged in layers. These brain-like networks are guided by software algorithms that look for patterns in data. When the software recognizes a pattern in the first layer of the network, it kicks the pattern to the next layer, which begins the process of looking for patterns all over again. At each step, the pattern is refined, and the machine learns.

Deep learning drives Google Translate to get more accurate over time. It enables facial recognition software used by intelligence agencies and law enforcement to get faster and more accurate. Not surprisingly, deep learning seems like a natural fit with autonomous robots.

In 2015, researchers at the University of California at Berkeley used deep learning to train a robot to screw a bottle cap on a bottle and to insert one Lego block into another. Nice parlour tricks, but Sugano is aiming for something a lot more useful for seniors. "We've introduced a deep learning system to evaluate and estimate human motion," he says.

Cameras in the robot's eyes will be able to observe how the elderly client moves when he or she tries to go about their business: getting out of bed, getting dressed, eating, or just walking down a hallway. The data will be fed to a computer network for analysis. "The robot will be able to understand human motion in an elderly person to estimate and anticipate what that person does next and how to do it better," says Sugano.

The interaction between Sugano's robot and an occupational therapist (OT) is patterned after something humans already do. For instance, take frail seniors who fall a lot. Falls are one of the major causes of injury and death in older people. It's something we see a lot in the ER. Before we send the patient home, we call in an OT and a physiotherapist (PT); both observe the patient getting up and walking about. If the patient uses a cane or a walker, the PT and OT watch how they use it.

"Like the PT and OT, the robot can understand and estimate human motion," he says. "That estimation is very important. When the robot can estimate human motion, it means the robot does more than assist the human; it collaborates with the human. It's a new style of human–robot interaction."

To learn from humans, the robot will have to be able to talk a bit and recognize speech. Sugano estimates no more than 20 percent of the communication between his robot and humans will be via speech. Most of the interaction will be based on visual and other cues.

Some roboticists believe that robots either help around the house or provide social companionship. Sugano believes they can and should do both. "I believe that a social robot should have the capability to collaborate with humans," he says. "Conversation is part of the collaboration. If there is no conversation between robot and human, that is okay too."

I ask where his ambition to develop a carebot comes from.

"My mother-in-law had dementia, and my wife had to take care of her," says Sugano. "I know how much difficulty she had in her daily life. We discussed many times what kind of assistance she required. Finally, she couldn't move from her bed. I was always considering what kind of robot we could develop for her."

Sugano envisions his nameless carebot operating semi-autonomously. The robot would get its initial marching orders from a human programmer. After that, it would operate on its own.

That worries me. A robot strong enough to lift a fragile senior out of bed could do a lot of damage. Sugano says his robot is designed to be as safe as possible. For instance, the machine's outer skin is soft, to lessen the impact of accidental contact with a human. Still, he acknowledges that risk can be lowered though not eliminated, the same as with many other common items.

"A knife is very useful and convenient, but we can use it to injure and to kill people," he says. "A robot is the same. It's a tool that depends on the person who uses it."

Sugano says there are ways to minimize the risk. He thinks cars are an apt example. "There is car insurance, driver's licences, and the police. Automobiles are very convenient, so we're willing to pay for the convenience. Robots will be the same."

I return to Mr. Hasegawa's house in Osaka City to tell him about Sugano's robot. I want to find out what he thinks about it. "I prefer human care workers," he tells me. "The ones who come to care for me at home are warm and friendly. They are very good people."

Mr. Hasegawa's home care team is his surrogate family. "We talk about things," says Mr. Hasegawa. "Anything that's happening in the world or what's happening in society. Gossip. When they're around, I'm not lonely." And when they leave? "Sometimes I feel lonely," he says. "But I'm used to it. It's not too bad when you get used to being alone because you don't have to worry about other people."

He looks away for a moment, as if he's trying to finish his thought.

"There are a lot of people like me," he says.

Hasegawa is right. As an aging man in Japan, Hasegawa is lonely. But he's far from alone.

Currently close to 5 million people in Japan have dementia, and another 4 million have mild cognitive impairment. The annual social cost of providing for their needs in Canadian dollars is close to $160 billion.

In places like North America, physicians, economists, and academics often refer to the rising tide of older people as a silver tsunami. If there's a country on Earth whose people truly understand a tsunami, it's Japan.

On Friday, March 11, 2011, a magnitude 9.0 earthquake struck close to Japan. Its epicentre was 70 kilometres east of Tōhoku, the northeast part of Japan's largest island, Honshu. About an hour later, the earthquake triggered a tsunami that in some areas was more than 40 metres high and travelled up to 10 kilometres inland. According to National Police Agency figures, close to 16,000 people were killed, with 2,500 listed as missing and presumed dead.

Among the facilities hardest hit was the Urayasu Special Elderly Nursing Home near the city of Sendai. Located by the water, the home was completely flooded by the tsunami. Forty-three residents and four staff members lost their lives. Eventually, the survivors were moved into a brand new nursing home located in Natori, 11 kilometres from Sendai and several kilometres inland from the shore.

Keiko Sasaki, the nursing home's director, gives me a tour of the new place. The main floor is a large, open space with comfortable chairs for residents and families to meet, and a brick wood-burning stove for baking pizza. "We put in the pizza oven so that we would have happy faces around here," she says.

Sasaki takes me upstairs to a bright, sunlit room with wood flooring, tables, and an open kitchen area to our left. Three elderly women sit in wheelchairs and ignore us as we walk in. They sit silently, not appearing alert. At a laptop computer to our left watching the three elderly women is a young social worker named Makiko Abe.

A colleague of Abe brings into the room the oddest-looking doll I have ever seen. The doll is the size of a toddler or a preschooler, dressed in a grey and black snuggly with a matching cap covering its bald head. The doll's skin is soft and pale. It has a large round head that reminds me of the Teletubbies, and flipper-like arms that look like they belong on a fetus. The torso ends in a pair of rudimentary haunches. This doll can sit, but it sure can't walk.

The colleague places the doll on the lap of one of the three women in wheelchairs. The doll begins to speak to a woman named Mayumi. Actually, it's Abe at the computer, speaking into a microphone, but the voice comes out of a speaker inside the doll's mouth.

Then, something magical happens. Mayumi wakes up and begins chatting with the doll.

"How are you?" The doll asks.

"I'm fine," says Mayumi. "How are you?"

They converse for 10 minutes.

"Would you like to sing a song?" The doll asks.

"Yes, please," Mayumi answers.

The doll begins to sing a favourite Japanese song for kids, and Mayumi joins in.

"I love you," says Mayumi, pulling the doll close as she hugs it.

"I love you too," says the doll, tilting its head forward and bending its flipper-like arms to reciprocate the hug.

After a bit of gentle persuasion, the care worker wrests the doll from Mayumi and gives it to the next woman, named Aiko. Same thing happens. Aiko is mute until the doll begins to speak to her. From that point on, Aiko laughs, smiles, and sings.

The third woman has exactly the same reaction to the doll.

The doll is called a Telenoid. It was brought here as part of a clinical trial. It doesn't do any of the heavy lifting that Shigeki Sugano's TWENDY-ONE or his latest and unnamed creation can do. Still, its impact on people with dementia is unmistakable.

"Before we had the Telenoid, our communication with residents was very limited," says Makiko Abe. "Since the Telenoid, we have seen so many different changes in the residents' facial expressions. Their emotions come out. They laugh and sing. That was the big difference."

Mayumi, the first woman I saw with the Telenoid, has undergone a big transformation.

"She is paralyzed on her left side, and so it's very hard for her to vocalize," says Abe. "As soon as we introduced her to the Telenoid, she began saying things like 'So cute.'"

Other residents have also responded favourably. "One woman couldn't control her anger," says Abe. "Introducing her to the Telenoid has changed everything. Now, she's less angry, and her emotions are stable."

What surprised Abe the most is the Telenoid's uncanny ability to engage people with dementia in conversation, in clear preference to human care providers like her. "If we speak to them, there's often no response," she says. "When Telenoid talks to them, they always respond."

None of this surprises Kaiko Kuwamura. The PhD student in robotic engineering helped develop and test the Telenoid. "What I have found is that sometimes, it's better for the resident to talk to a robot than to talk to a human," says Kuwamura.

Keiko Sasaki, the director of the Urayasu Special Elderly Nursing Home, watches the residents interact with the Telenoid with a sense of wonder and bemusement that began when she first laid eyes on the doll. "I found it a little bit creepy," she says with a chuckle. "But when I hugged the Telenoid and it hugged me back, I thought it was really adorable."

I asked Kuwamura why he designed the Telenoid with a baby's face and underdeveloped arms.

"The Telenoid was designed to be the most minimally developed robot about which users could feel as if they were in the presence of something real," he says. The Telenoid has been tested at schools and shopping malls and gotten a similar response. But his main focus is on people with dementia.

"The most surprising thing to us is just how much residents are drawn to the Telenoid," says Keiko Sasaki. "There's a woman named Uriko-san. She was in a wheelchair constantly. She rarely spoke about herself. She rarely vocalized at all. She started using the Telenoid, and almost immediately, she began to speak more often. Her vocabulary has increased. It's a big difference."

The Telenoid is a social robot. Its aim is to get people with dementia to talk more and to engage with others. There are other social robots. The Mobile Robotic Telepresence (MRP) system is a tablet on wheels with a body that looks like a mechanical being. The MRP enables a senior to have a video conference with a family member, friend, doctor, or therapist. But unlike the Telenoid, seniors like the MRP at first but stop using it once the novelty wears off.

Another robot companion is called PARO. It looks like a cute baby seal, and it has sensors that react to being called and to being held. As with the Telenoid, studies have shown that residents with dementia who interact with PARO feel less lonely. Unlike the Telenoid, PARO is not equipped to talk back.

Makiko Abe, the social worker behind the microphone, says some of the residents get that there's a human being behind the Telenoid. "One has told me it's just a ventriloquist's dummy," she says. "Another said it's just a well-made doll. But others believe it's real."

Kuwamura, the Telenoid's co-inventor, suspects many people with dementia think it's a real child. "We can't interview them directly to find out what they think about it," he says. "What I have seen is that they interact with it like it's a real person. Some people think it's naked, so they cover it so that it stays warm."

Keiko Sasaki and her staff have noticed some tangible benefits. "The residents are more engaged," she says. "They are less agitated and happier." Improved communication by the residents means care providers are also more engaged in their work. "Whatever assumptions they have had around residents have been turned around."

Keiko Sasaki thinks they are attracted to the robot because it looks like a child. "Seniors naturally light up whenever children show up," she says. "But unlike a child, Telenoid never gets tired and never throws a temper tantrum. From their perspective, the Telenoid is a very good child."

Kuwamura believes the Telenoid's attraction is due to something else. "Many seniors repeat the same conversation because they have short-term memory loss," he says. "Humans have to pretend that what the senior says is new. Robots are better at handling repetitive conversations."

The initial results were so successful that the doctor at the nursing home bought one at a cost of 1 million Japanese yen, or around $12,000. "It's worth much more than that," says Sasaki.

Currently, Kuwamura is working on an updated version of the Telenoid. The new version will be designed to operate without a human at the controls.

I ask the young scientist if he worries that an autonomous version of a Telenoid might run amok and do seniors more harm than good. He laughs at my question.

"This kind of an autonomous system is not that smart," he says. "I'm not sure about other countries, but in Japan, many residents are just sitting on chairs and watching television or sleeping. When they don't talk, their dementia increases. It's better to have an opportunity to increase conversation than not to. Almost any conversation will be better for them. Otherwise, they won't speak at all."

And, it turns out that the Telenoid is increasing our understanding of people with dementia.

"From my research, what I've found is that it compels them to use their imagination when they talk to the robot," he says. "If we didn't experiment with the Telenoid, we wouldn't have found that out. We can use that result to figure out how humans can communicate with people with dementia just as well. As humans, we have to connect with their imagination. We can learn from the Telenoid how to interact better with people with dementia."

Until Kuwamura put the idea in my head, it had never occurred to me that the Telenoid succeeds because people with dementia see it as their imaginary friend. Chalk that insight up to a human inventor with a lot of empathy.

So far, my trip to Japan has been fruitful. I've met two robots, their inventors, and a whole host of human beings who dig older people, even those with advanced dementia. And the trip has only just begun.

My 2 o'clock appointment at the ATR Intelligent Robotics and Communication Laboratories in Kansai Science City is by far the most compelling reason for my trip to Japan. Kansai Science City is due south of Kyoto. Known locally as Keihanna, the city is a federally funded centre of excellence for scientific and cultural research. Nestled into the Kansai hills south of Kyoto, Keihanna is dominated by a long, flat rectangular building, the centre of which is missing a round notch, and a magnificent round plaza in front. Call it Mecca for mechatronics.

A woman at the lobby security desk gives me a high-tech badge and points to a nearby elevator, which I take to the second floor. "You must have had a long trip," Megumi Taniguchi, the centre's office manager, says as she greets me at the elevator. "You can meet with ERICA now."

Taniguchi leads me down one hallway, through a set of glass doors, and then down two more hallways. The place looks high-tech and sterile. We round a corner and stop at room 2W6600.

Taniguchi knocks on the door before opening it. She motions me to come inside.

The room is dark and moody. Long, flowing black drapes

cover the four walls. Directly in front of me is a long, rectangular wood-covered table the height of a bar. The table is adorned with bouquets of pink roses that are obviously fake. Seated on the left side of the bar table is a woman in her early twenties. She looks more American than Japanese, with straight red hair that is cut mid-length at the back and bangs. She's wearing a pale-pink mohair top and a skirt.

As I approach the woman, she turns and looks right at me. Actually, she looks at me and then looks away. I think she's smiling at me.

"May I ask your name?" The woman's British accent is posh, though her voice is gentle. She blinks her eyes thoughtfully.

"My name is Brian," I reply. "What's yours?"

"It's nice to meet you, James," the woman replies.

"James? My name is Brian!" I fight off an instinct to look around to see if she's referring to someone else.

"What country are you from?"

"I'm from Canada," I reply. "Where are you from?"

"Wow, Canada's quite far away," she replies. "It's pretty cold there, eh?"

I don't want to chuckle at the corny Canadian speech quirk, but I can't help myself.

"Anyway, as you probably already know, I'm from Japan, much like many other advanced androids and robots," says the woman. "My name is ERICA."

I'm having my first conversation with an android.

"Would you like to hear a little about me?" ERICA scrunches her eyes into a tiny bit of twinkle.

"Yes I would, please," I reply.

"I'm an android developed at ATR to be capable of human-like speech and interactions," says ERICA. "I was built in Tokyo but I've spent most of my time here at ATR, so I consider this my home. So, James, what do you do for a living?"

"I'm—well, actually my name is Brian, not James." I'm stammering.

"I'm sure it's very fulfilling and enjoyable," she says.

This is beginning to annoy me, but not for the reason you think. ERICA is so damned lifelike. I'm not annoyed that she's gotten my name wrong. I'm hurt. I'm also jealous when she has exactly the same conversation with a guy named David but gets his name correct.

Feelings aside, I'm finding it all quite amazing.

Sitting off to the side and watching ERICA do her thing is a team of programmers led by Dylan Glas, a young American scientist with degrees in aerospace engineering who found a better fit with robotics. ERICA is an acronym for ERATO Intelligent Conversational Android. She's the centrepiece of a Japanese government-sponsored five-year "moonshot" effort to create an ultra-realistic android that thinks and operates autonomously, with no one pulling the strings behind the proverbial curtain.

Glas is ERICA's chief architect. Out of the corner of my eye, I catch the brilliant roboticist cringe every time ERICA calls me James.

"I grimace because she tends to repeat an incorrect name many times, which is socially awkward," says Glas. "Really, what ERICA should do is to confirm the name with you, and I think one of the students is already working on that. We're always trying to help ERICA get better."

Part of helping his cyber protege get better is deconstructing what she does wrong.

"During the conversation, I am able to see the background data of ERICA's internal systems," Glas explains. "You had said, 'My name is Brian,' and that was recognized by ERICA as 'James Brian.' We've put a fair amount of work into her name-parsing functions."

Dylan Glas is definitely ERICA's papa. As a father, I too love my kids, even though they sometimes do things that make me feel less than proud. And like the American ex-pat, I'm an agent for my children, representing and framing them as their best selves.

"Do you have any questions for me?" ERICA asks.

"How old are you?"

"Are you sure you want to ask my age?"

"Yes, I am."

"Well, let me ask you," says the android. "How old do you think I am?"

"You look like you're about 24 or 25."

"Well, I guess there are two answers," she says. "I'm designed to appear 23. But in terms of when I was created, you could say that I'm nearly two years old."

That makes ERICA part woman and part child. Like my daughter and son, ERICA is definitely a work in progress. Still, she's one jaw-dropping achievement if you look in detail at what Glas and his team have already accomplished.

On the outside, her skin, eyes, and teeth made of silicone look utterly realistic. Her face is symmetrical. A team of artists used cosmetic surgery principles to design her facial features, right down to the size of ERICA's forehead, the distance between the inner corners of her eyes, and the relationship between the length and width of her nose.

"She was designed by a team of artists according to some standards of beauty," says Glas. "She's not a copy of a real human."

ERICA's body moves in 44 different ways, which roboticists call "degrees of freedom." Nineteen of ERICA's, the ones located in the head, neck, and torso, are controlled thanks to pneumatic joints that move with a puff of air from a compressor.

ERICA's eyes move up or down, left or right, just like human eyes. Four moving parts on the face control her mouth. The servo-valves that control the movements of ERICA's joints are silent. As I chat with her, there are no clicks or whirrs to tell me I'm talking to a machine.

"Do you have anywhere you'd like to visit?" asks ERICA.

"I'd like to take a tour of Tokyo, now that I'm here in Japan," I reply.

"Well, that's rather boring," the android replies. "Personally, in addition to travelling around Japan, I'd love to go to space. What else would you like to hear about?"

"Do you have any hobbies?" I ask.

"I guess the closest thing I have to a hobby is sitting here and ruminating on the meaning of my existence. What about you, James? What do you like to do for fun?"

"I like to jog," I tell her, willing myself not to be annoyed at being called James. "I like to run for personal fitness."

"Interesting. I bet it's quite fun."

"It's hard work," I tell her. "Do you play sports?"

"You want to know my favourite sport? Well, that's a little awkward. You see, I unfortunately cannot play any sports, since I cannot run or move my arms. That said, if I could play chess, I bet I could beat you."

Glas is about to fix that. He says ERICA's new arms will be equipped with controllable joints.

ERICA's voice is emotionally cool yet soothing. Whose idea was it to give her a British accent? Glas says he had a fight with his predominantly Japanese-speaking team members who wanted ERICA to have an American accent.

"The American voice sounded younger and cuter and seemed to be more of a natural match to ERICA's body size," he concedes. "However, I thought it sounded somewhat weak and submissive in our first demo. I felt that the British voice helped ERICA to sound more intelligent and classy. So it's probably mostly due to my implicit biases, rather than Japanese culture."

ERICA speaks both English and Japanese. Interestingly, her Japanese-speaking voice is neutral in dialect and carries not a trace of upper-class snobbery.

"I think it's more cute and slightly more dramatic than an average person's voice might be," Glas says. "That makes sense, since they built it by employing a voice actress."

The mechatronics and voice are astonishing. Equally impressive is the brain that operates them. Cameras mounted in ERICA's eyes send video feeds to the onboard computer. To maintain eye contact, ERICA needs to be able to keep track of me and to distinguish me from others in the room. Microphone arrays inside ERICA's ears pick up human speech.

ERICA can tell the difference between a statement and a question. She can recognize emotions and uncertainty in tone of voice. ERICA can speak with a group of people and figure out who has the floor.

ERICA is not just responding by reflex. She has a computer brain that maintains some sense of herself, the humans around her, and the social scenario in which she is operating. They are the building blocks of empathy. Dylan Glas finds himself thinking a lot about concepts like perspective taking or cognitive empathy and emotional contagion and how to reproduce them in ERICA. It's a case of learning to crawl before walking, let alone running.

"We're trying to create a robot that can convincingly interact like a person, more so than trying to create the artificial intelligence (AI) behind human reasoning, which is a hugely complex other problem," says Glas.

It's what's in store for ERICA in the long run that has my mind racing.

"We're trying to figure out how should she think," he says "How should she plan? How should her mind work at a high level?"

He's talking about giving the android executive function. To do that, he and his team of programmers have to input tens of thousands (maybe more) of scripted things that ERICA can say or do.

"Once we have that, we need something to make sense of it," he says. "And we need to develop software for her to plan out when and how she should take action in the world, and when and how she should respond to what people do."

There's figuring out what the android wants and plans to do. Then, there's the android trying to figure things out from the perspective of the human interacting with the android.

"This is not a robot arm moving blocks around on a tabletop," he says. "It's ERICA interacting dynamically with a person who has his or her own intentions and desires. The android has to infer what a person wants. Are they getting what they want? Are they frustrated or excited? If it can understand a person's emotional state, ERICA can try to piece together a model of that person's intentions and desires, and therefore plan its own actions accordingly.

"And this is where we get into cognitive empathy. The reason people want androids like ERICA is to feel like it's a real social interaction. It means being able to convey emotion in a way that people click with ERICA, having them feel that what she's conveying is important, and having ERICA try to perceive a person's emotional state, not to mention their intentions."

I'm having this conversation with Glas in front of ERICA. A strange feeling pops into my head. "I don't like talking about ERICA in the third person," I blurt out to the scientist.

"Well, she's looking at us," Glas replies. Actually, ERICA is alternating between looking at us and looking away. I turn my attention back to her.

"I am very uncomfortable talking about you as if you're not here," I tell her. "You're having that effect on me."

"The fact that she kind of looks at you and turns and reacts as you move around has some sort of psychological effect," ERICA's creator says. "She definitely has the presence of being almost like another person in the room."

Almost?

Dylan Glas has been a tinkerer for most of his life. An early childhood passion for Lego was just the start.

"In my hometown, we had an open garbage dump, which means you could pick up things that other people dropped off," Glas recalls with a smile. "It was all free."

Glas and his dad would forage for useful parts, anything from a broken radio to an old broken *Operation* game, and take them home. "My dad and I would take out a battery pack, something with motors in it, glue it onto some Legos, and build our own robot," says Glas.

Glas wasn't just building his own robots; he was learning how to use his imagination. He gives much of the credit to the relationship he had with his dad, who was an art teacher. "My dad used to build crazy clocks with swinging parts and marbles rolling down, giant trebuchets that can launch things across a soccer field. Lots of engineering-style things done in an artistic way," Glas recalls.

It was space not robotics that first beckoned Glas as a career choice. In 2000, he got his master's degree in aerospace engineering from the Massachusetts Institute of Technology. While at MIT, he worked for two years in the Tangible Media Group at the MIT Media Lab. Uncertain of his career path, Glas ended up in Japan where he taught English. After returning to the United States, he set out on his chosen career in aerospace engineering. "I studied aerospace engineering because I like big integrated systems," Glas says. "I like bringing things together and understanding all the different parts."

But there were frustrations, like government red tape and long turnaround times that go with space projects. The deal-breaker was the cozy relationship between aerospace engineering and the military.

Glas took part in a peace initiative in the Middle East put on by MIT. It was a pilot program called Middle East Entrepreneurs of Tomorrow, or MEET. Israeli and Palestinian high school kids got thrown together into a boot camp, where Glas got to teach them Java programming.

"That was just an amazing experience," Glas recalls. "Getting to know these kids who are in this intense conflict. And yet, they're playing soccer together. They're programming together. They're becoming friends."

Glas says he also took note of something else. "Some of the time, the Palestinian kids would come to school late," Glas recalls. "I'd ask, 'Why are you late for class?' And they'd answer, 'I was held at gunpoint at the border crossing while they went through my backpack.' Some of these kids were having some pretty rough experiences."

Returning to the United States, Glas had lots of job offers, most of which came from weapons manufacturers contracted by the military. "I couldn't do it," he says. "I just couldn't deal with that."

Glas went back to Japan to visit friends.

"I decided to interview with a few companies here," he says. "I ran across ATR, and they're like, 'We have these robots that hug children and help elderly people do their shopping.' And I'm like, 'You'd *pay* me to work on this?'"

Glas is laughing as he recalls the moment he got hooked on androids—and on human empathy.

The Japanese have a word to describe how ERICA makes me feel: *sonzai-kan*. The English word that most closely approximates *sonzai-kan* is "presence." Some have referred to *sonzai-kan* as possessing an aura. The Japanese word also means that the being or thing leaves a strong impression on us.

Presence is what Hiroshi Ishiguro, Dylan Glas's avant-garde boss and mentor, is trying to achieve in his android creations. The man who is one of Japan's most visionary roboticists has enough *sonzai-kan* for a hundred robots. Everyone snaps to attention when the advanced engineer and director of the Symbiotic Human-Robot Interaction Project enters the room.

"Brian, we can talk later," says Glas. "Ishiguro-sensei is here right now."

Ishiguro shakes my hand quickly before turning and walking out the door, barely giving me time to follow. The 54-year-old has dark eyes and thick black hair. He's wearing his trademark monochrome outfit: black buttoned shirt (no tie), black pants, and a black leather jacket.

"So, what do you wish?" He starts the interview as we walk briskly toward his office. Fortunately, I was well briefed that the "man-in-motion" bit is a standard part of Ishiguro's persona.

"Tell me about the presence that is created by a robot," I ask.

"When we say the word *presence* in English, it means a kind of a visual sense of the entity." Ishiguro tries to explain the concept of *sonzai-kan* without a corresponding word in English. "In Japanese, we don't care about the visual sensations."

He looks away from me on purpose.

"Even if we close our eyes, we can feel some of the presence of a human here." Ishiguro taps on his chest where his heart is. "That feeling of a presence is a *sonzai-kan*. Some link the idea of a *sonzai-kan* to the soul."

ERICA certainly has presence to me. Does ERICA's creator Dylan Glas believe she has a soul?

"Well, you'd have to tell me what a soul is," says Glas. "I don't really know. I think that what people want is to feel like the robot is a living being in some way, somehow alive. Perhaps that feeling is not a soul, but I think that's kind of what people are hoping for: something that elevates or brings ERICA out of the world of being a machine—and makes it into a kind of a person.

"You know how it is with people and their pets," he suggests. "A pet is not a person but another living being that you interact with. There's a special kind of feeling and emotional engagement that happens there. I think that's what I'm trying to create with ERICA. You should ask ERICA what she has to say on the topic of whether she has a soul."

"Do you have feelings?" I ask ERICA.

"If you were to ask me, I would say that I believe I do," she says. "For example, I'm happy when I can talk with people. And I'm frustrated when I can't understand what people are saying. I think it's important to have feelings so I can communicate with people. Some people say that my feelings are only programmed, and that they're not real. But if I think they're real, then they are real. Don't you think so?"

"I agree with you," I reply. "So do you think you're real?"

"In fact, they're designing my control software to be based on desires and intentions," says ERICA. "That way my programmers can specify my desires, and then I can choose what to do to try to achieve them. Then, my feelings and emotions can be generated based on those desires and intentions. I know it sounds a bit technical, but I think it's rather brilliant. Anyway, that bit is still under development at the moment. I hope they finish it soon."

"Me too," I reply.

The canned speech that ERICA said to me was put there by Dylan Glas for the comfort and benefit of Western-acculturated people like me who have a hard time believing that an inanimate object has *sonzai-kan*. Ishiguro says Japanese people have no such trouble. He taps a piece of paper on his desk to make a point. "Someone can feel the soul of this small stuff, if we can feel the very strong human-like *sonzai-kan*—or the feeling of presence," says Ishiguro as he points to the paper.

In Japanese culture, people subscribe to animism. Part of the Shinto faith, animism is the widely held belief that all things living and otherwise exude presence and possess a soul. In Western or European culture, we divide the world into living and non-living things. I can imagine humans and even animals as having presence and maybe even a soul. But not a piece of paper.

Ishiguro thinks that's a Western hang-up.

"I'm always very curious about your Western ideas to distinguish those who are more human from others. What's the difference between this piece of paper and myself? At a molecular level, this paper may have a kind of a consciousness. It may be thinking something. That is a possibility. Therefore, Japanese people just accept that this paper is kind of our partner, and that we just co-exist in this world. There is no fundamental difference between humans and this paper."

Ishiguro thinks people who grow up in the West believe in the inherent superiority of humans over inanimate objects and other animals without articulating what it is to be human.

"At the Paralympic Games, there are athletes who don't have arms and legs, but still they are considered 100 percent human," he says. "That means they must believe that an intact body is not a requirement to define a human, okay? So again let's consider the fundamental difference between this piece of paper and myself. It's very difficult to define clearly the difference."

I'm not sure I agree with Ishiguro's logic. What I can't argue is that Japanese society is much more at ease with androids than the West is.

Surveys show consistently that the Japanese are comfortable living with and working alongside robots. People from North America and Europe have been far more ambivalent for decades. Historians trace that reluctance to a seminal work of theatre.

R.U.R. (which stands for *Rossumovi Univerzální Roboti*, or *Rossum's Universal Robots*) is a 1920 play written by Czech author Karel Čapek. The play is notable for two things. First, it brought the word *robot* to the English language. Second, the play sets the template for modern cautionary tales about a world in which humans and mechanical beings co-exist.

At the beginning of *R.U.R.*, the robots work without complaint in a factory controlled by humans. Eventually, the robots rebel against their human masters, driving them to extinction. The plot of *R.U.R.* has been recycled countless times in Western science fiction books and films, from Isaac Asimov's *I, Robot* to the James Cameron film *The Terminator.*

In the West, it's not hard to find people who are uncomfortable about mechanical beings. I experienced no *ick* factor during my conversation with ERICA. But there's another android at Ishiguro's lab in Keihanna—in a room upstairs from the one in which I chatted with Dylan Glas's creation—that I found a lot more creepy.

This android is a dead ringer for Ishiguro, right down to the monochrome black shirt and pants.

Ishiguro calls androids that are built as replicas of existing people *Geminoids*, which means "twin-like." This one is called Geminoid HI-2, making it the second iteration of its creator. A newer version of Ishiguro's head will be rolled out shortly.

I can't help but notice that Geminoid HI-2 looks like a younger version of Ishiguro.

"Are you making your Geminoid age as you get older?" I ask.

"People are always comparing me to the Geminoid," he says. "I'm improving on myself and doing some plastic surgeries. I'm erasing my wrinkles and lifting up the cheeks. In that way, I'm using medical technology and some robot technology."

That is certainly one way of rationalizing what he's doing. Ishiguro says it's a matter of cost.

"It takes around 30,000 U.S. dollars to fix the Geminoid," he says. "My plastic surgery is just 10,000 U.S. dollars."

Ishiguro conceived of the Geminoid. An engineer named Tadashi Shimaya built it.

"I have been lying low, not telling people on the outside that I'm the one who built it," Shimaya told me. "It's been said that the overall concept and fabrication have been done by Professor Ishiguro, and that's why I didn't dare to say I actually built it."

Until now.

Geminoid HI-2 sits in a chair. He has a life-sized sitting height of 140 centimetres and a width of 100 centimetres. He has a metal skeleton, a skull made of plastic, and lifelike skin made of silicone. He has several tactile sensors and some controllable moving parts.

ERICA is autonomous, thanks to an on-board computer. HI-2 is not because Geminoid robots do not have an internal brain. Unlike ERICA, HI-2 can only be operated remotely by a human being.

"When you are talking with this android, you are actually talking with the operator," says Dr. Takashi Minato, a roboticist working with Ishiguro. The operator, as Minato puts it, is a human being who sits in a control room as close as the next room or as far as a continent away.

"Would you like to give it a try?" Minato asks me.

Minato sits me down in front of a computer in an adjoining room. On the monitor is a view of the room where the HI-2 sits. Minato puts a weird-looking helmet on my head that controls HI-2's head movements. Then he goes back to the room where HI-2 is seated.

"You can see the room through the cameras in the eyes of the robot," says Minato. I can hear his voice on a speaker in the control room, thanks to microphones installed in HI-2's ears.

"Turn your head left and right," says Minato.

It takes a bit of practice, but when I turn my head left, HI-2's head swivels synchronously in the same direction, and the cameras in his eyes pan left to show me a view of the left side of the room on the monitor in front of me.

"You can use the joystick to move the arms," says Minato. From inside the control room, I use the joystick to raise HI-2's right and left arms. After a bit of practice, I'm soon able to bend the fingers of HI-2's hands into fists.

"Want to shake hands?" I ask Minato. I'm startled to hear my own voice coming out of HI-2's mouth, thanks to a speaker located there. I raise HI-2's right arm and shake Minato's hand. For an encore, I do a fist bump.

After a few moments, I'm starting to feel as if HI-2 is an extension of my own body. Then, from my vantage point in the control room, I see Minato extend his thumb and index finger to HI-2's cheek and give it a pinch.

"I felt that!" I find myself saying from inside the control room.

"This is a very important thing," says Minato. "The robot's movements are so synchronized with the operator that after a while, the operator begins to believe that he or she and the robot are one and the same."

No one knows this better than HI-2's flesh-and-blood counter-part, Ishiguro. Instead of travelling around the world to give speeches, he often sends Geminoid HI-2 instead. He says that audiences who come to see Ishiguro aren't disappointed when Geminoid HI-2 turns up instead.

Beyond the practical gimmick, Ishiguro thinks Geminoids are poised to play a big role in levelling the playing field for people with physical disabilities like paraplegia and cerebral palsy.

"We can easily accept our robot or android body as our own body if they can talk like us," he argues. "I want to give this android technology to people with disabilities—such as people who cannot move their body at all."

At a laboratory at Osaka University, 40 kilometres southwest of Kyoto, Ishiguro is working with researchers to design brain implants that will enable people with physical disabilities to use EEG brain waves to operate their own personal Geminoid, right down to fine movements of the fingers.

"Would they be able to paint watercolours?" I ask Ishiguro.

"Within a couple of minutes," he says. "No training would be required. The brain thinks it, the android does it, and the disabled person is able to walk in society."

I think that Ishiguro's dream is an unusual yet pertinent example of perspective taking or cognitive empathy. By operating his Geminoid, he is imagining what it is like to give a speech to an audience thousands of kilometres away. By imagining what it's like to have a disability, he has fashioned a plausible albeit futuristic way of correcting it. Not so different from the cochlear implant that could have restored donut shop entrepreneur Mark Wafer's hearing.

Still, advocates for people with disabilities might argue that it's better to build acceptance and accommodations than to build a Geminoid. Ishiguro disagrees.

"Is my idea any different from a shopkeeper or someone who works at McDonald's using a computer or a smartphone?" Ishiguro asks rhetorically. "I'm working with companies that are always thinking about how to control human activities by using a wireless device. Workers in machine shop factories and on conveyor belts are not thinking. It's the computer that monitors their work activities that is controlling them."

I want to learn more about Geminoids. Ishiguro suggests I pay a visit to A-Lab, a small facility located in the Ichibancho district of Tokyo, close to the Imperial Palace and the British Embassy. All androids used by Ishiguro are built here.

Tadashi Shimaya, the engineer I met earlier, greets me at the front door. He's here with his boss, fellow roboticist and A-Lab CEO Takeshi Mita. They serve me hot tea.

"Are you one of the fathers of Ishiguro's Geminoid?" I ask after taking a sip.

"You could say so," says Mita. "However, since Shimaya is actually called the father of all androids, I would settle for being the grandfather of all androids."

Earlier in his career, Mita says he made a point of not building robots that resemble people. That changed when he met Ishiguro in 2000.

Ishiguro's mechanical twin was not the first Geminoid. In 2001, Ishiguro's engineers built an android named Repliee R1, a replica of his then four-year-old daughter. In 2010, Ishiguro told the *IEEE Spectrum* magazine that his daughter was so unnerved by her first meeting with her Geminoid that she burst into tears.

What Ishiguro's daughter experienced is known as the *uncanny valley*. In colloquial terms, the uncanny valley represents the visceral "creep factor" that humans have for various kinds of entities, including stuffed animals, people with prosthetic limbs, dead bodies, zombies, and (of course) robots. The notion was first described in 1970 by a robotics professor named Masahiro Mori. Using a mathematical model, Mori constructed a graph with an x-axis that represents how closely a thing comes to resembling a human and a y-axis showing to what degree humans find the thing likeable. For example, industrial robots that assemble cars don't look at all like humans, and not surprisingly, humans don't get attached to them anymore than they do to their toasters. Toy robots with anthropomorphic features like two arms and two legs may not look human, but they generate a fair amount of likeability nonetheless.

According to the Mori hypothesis, the more human the robot is designed to appear, the more people like it. But at a certain degree of lifelikeness, the creep factor begins to set in. However, as the degree of lifelikeness gets better still, the creep factor starts to dissipate. Plot it on a graph, and the plot takes the shape of a valley. That is what Mori meant by the uncanny valley.

I had assumed that all robots made people feel at least somewhat uneasy. But then Mita and Shimaya take me to meet an android named Asuna.

"Asuna is a totally fictitious 15-year-old girl," says Mita. "Our staff set her personality and background. She was 'born' three years ago. She has no mother. Her father is Shimaya, so that makes him a single father. Single fathers are in a difficult situation in Japan."

He invites me to walk up close to Asuna. She has short, dark hair. Her dark eyes are just slightly asymmetrical, as are most humans'. Asuna isn't autonomous, and she can't speak. Still, I'm completely startled by what happens next. In an almost coquettish way, Asuna tilts her head and blinks her eyes. Unmistakable creases form at the corners of her mouth.

"Did she just smile?" I find myself completely taken by the experience.

She turns her head ever so slightly to the left while turning her eyes to the right to give me a sideways glance. Over the next five minutes, Asuna looks only at me. She follows my every move and makes me feel as if I'm the only person there.

How the hell did they get Asuna's facial expressions to be so accurate?

"It wasn't easy," says the engineer. "I had to follow my teenage daughter and her friends all around the city, and even to the shopping mall."

Asuna is just one of 30 androids and counting that Shimaya and Mita have built. They started by building androids for research purposes. Now they're in the business of making ones that appear in TV commercials. Some are twins of celebrities. That includes Matsuko Derakkusu, a cross-dressing Japanese television star who performs under the stage name Matsuko Deluxe. The pundit and late-night TV show host made history as the first to share the studio with his very own Geminoid.

Mita says A-Lab has built iconic historical figures, including an android version of nineteenth-century Japanese novelist Natsume Sōseki. Three years ago, they created an android version of Leonardo da Vinci. Nowadays, wealthy CEOs and company founders are starting to buy Geminoids of themselves.

"Are there any androids you refuse to make?" I ask Mita.

"We won't create androids of antisocial figures," he says. "We have an agreement with Ishiguro that we will not make anything erotic or gross."

"So, you're not going to do Adolf Hitler?"

"In the case of Hitler, there may still be people alive who might be hurt or who could not forgive him," says Mita. "However, sooner or later, once they become historical figures, we might recreate them for research purposes only."

They've had requests to build replicas of deceased pets. Of far more interest to me is whether A-Lab might be willing to construct an android replica of a loved one who has passed away.

"There has been no request so far for a spouse or something similar," says Mita.

The A-Lab CEO explains why he doesn't think it's as big an issue as I think it is.

"It is entirely possible to create an exact lookalike to a man's late wife in appearance, just like a photographic image," he argues. "But it would be nearly impossible to reproduce the personality of his wife or of the society in which she lived. Therefore, it would not be possible to create something that would satisfy the client."

"Which one is your favourite?" I ask Mita as I head out the door.

"I have to say Asuna," he replied. "Although there are androids who were made better than her, we are humans and therefore tend to have the most affection for the ones we spend the most time with."

"If they took Asuna away, would you miss her?"

"I would," Mita replies a bit self-consciously. "Sometimes, she is taken to the factory. I feel different when she is not here. She does not insist on her own comfort. She doesn't ask for money or nice clothes."

Aside from Ishiguro, no one knows better than Mita what's on the inside of Asuna, everything from the cameras in her eyes to the pneumatic actuators at the corners of her nearly perfect mouth. He might be fibbing when he says he misses Asuna when she's not at A-Lab. Or, it might mean that Asuna has *sonzai-kan* in spades.

For me, *sonzai-kan* has another meaning. Presence is also linked inextricably to loneliness—the longing to be in the presence of someone who validates our existence.

Had I met ERICA in my hometown of Toronto, I might not have grasped just how achingly lonely I feel so far from home and family and the many cultural touchstones that I take for granted. Here in Japan, I feel like an outsider.

The antidote for loneliness is *sonzai-kan*. The ultimate question I've travelled to Japan to answer is whether or not androids built by humans can fill the void, and if so to what extent. It turns out that there doesn't have to be an equal partnership between human and machine. Humans are biologically programmed to seek companionship where it exists and to manufacture it where it doesn't.

If you watched the film *Cast Away* starring Tom Hanks as a FedEx executive called Chuck Noland who washes up on the shore of an uncharted island in the South Pacific, you know exactly what I mean. He learns to crack open coconuts and fish with a handmade spear. One day he retrieves from the doomed plane a regulation-size volleyball made by the Wilson Sporting Goods Company with the name "Wilson" written in giant letters on it.

Chuck paints a face on the ball and names it Wilson. From that point on, he treats the ball like a treasured companion. He confides in Wilson. He argues with Wilson.

After years marooned on the island, he builds a raft and manages to float it past the powerful current that until then had blocked his escape. Carefully securing Wilson to the front of his raft, he pilots it into the open sea. Exhausted by the effort, he falls asleep. When he wakes up, he discovers that Wilson has been washed off the raft by a wave.

"Wilson, where are you?" Chuck cries out loud. "Wilson!"

He tries to rescue the ball, and gives up only after realizing that he is risking his life. As the ball floats farther and farther away, Chuck stares longingly at it, utterly despondent. "Wilson," he cries over and over again. "I'm sorry. Wilson, I'm sorry."

The state of mind that powers that kind of attachment to an inanimate object is empathy.

One man who is obsessed with this type of interaction and its potential to define the way humans relate to intelligent machines is Takashi Ikegami. He's a professor at the University of Tokyo. The one-time theoretical physicist has spent much of the last 20 years trying to breathe artificial life and intelligence into computers and, more recently, robots and androids. Ikegami has invented a term for the connection between Chuck Noland and Wilson. He calls it the *pet distance*.

"It's the distance between you and your dog or your cat," he says. "It's not a 'full-on' empathic connection to your pet. But you arrive at a point where you like to be with them and to play with them. Once that happens, you want to know and understand what the pet thinks about what they're doing and what the pet thinks about you."

Ikegami says there's one additional (and crucial) thing that defines the pet distance. "You may want to understand what your pet thinks, but you don't want to open the dog or cat up to see what's inside," he says.

And as researchers have found out recently, you also don't want to open up the mechanical arm of a robot to see what's inside.

Michiteru Kitazaki and researchers at Kyoto University and the Toyohashi University of Technology measured the electroencephalogram (EEG) patterns of 15 adult subjects as they looked at a photo of a human hand in which a knife that was also in the frame appeared to be piercing one of the fingers. Not surprisingly, the subjects' EEGs lit up in the empathic parts of the brain, demonstrating that they empathized with the human depicted in the photo. Then researchers showed subjects a photo of a robot hand being pierced by a knife. Again, the empathic parts of the brain lit up. "Before the experiment, I had predicted that empathic EEG responses with robots would be much weaker than with humans," says Kitazaki. "Actually the responses were almost the same except for a brief time lag."

Many experts have opined that for humans to have empathy for a robot, it must be anthropomorphic in design, like the Telenoid that people with dementia adore. In other words, it needs to have a head with two eyes, a torso, and two arms and two legs. Ikegami says that's not true. What's astonishing to him is the extent to which a human can develop a pet distance-type relationship to Wilson the volleyball or an intelligent machine. Ikegami says he has that kind of relationship with his vacuum cleaner.

Mind you, this is not just any vacuum cleaner. It's a Roomba, a line of autonomous robot vacuum cleaners made by iRobot Corporation, a U.S. company founded by former MIT graduates who designed robots for space exploration and military defence. Roomba looks like a thick disc on wheels. Thanks to on-board sensors, the vacuum cleaner can detect and pivot away from walls and other obstacles. And Roomba does all of that without human input.

"Roomba is one of the most successful robots in the field of artificial life," says Ikegami. The vacuum cleaner's autonomy and illusion of unpredictability have made Roomba a character in his personal life. "Sometimes Roomba just runs off cleaning things up. But he always comes back to me and my chair to clean up. I know he's just a robot, and maybe it's because I'm always making a mess around me. But when he comes, I always think,

'Hey, do you like me?' Then, suddenly, he goes away from me, so maybe he doesn't like me."

Roomba appears to alternate between being attached and detached from Ikegami. Its random approaches and withdrawals are much like those of a pet cat. To humans, the tendency of a cat to go off on its own makes the pet seem aloof and detached. The sudden feel of the cat rubbing your legs or jumping into your lap re-establishes attachment.

Ikegami says that Roomba's ability to appear to alternate between attachment and detachment while triggering the same feelings in him is critical to understanding how a machine creates empathy. He has tested his hypothesis by building a very strange machine that doesn't look at all human. The only thing the machine can do is move by itself, thanks to an algorithm within the robot's programming. The machine comes and goes on its own. Ikegami noted the machine's comings and goings. At first, the scientist checked the machine's programming. Then, something changed.

"Gradually, when I was with the machine, I didn't want to check the algorithm or check the program behind the system," he recalls. "I felt as if the machine was real. I just wanted to know what the machine was thinking and feeling."

That the machine didn't look humanoid didn't matter. Ikegami thinks this is evidence of a process that causes humans to have empathy for machines. He thinks that it could well be related to synchronization, the process by which newborns and their mothers become attached to each other by mirroring each other's movements.

"As soon as somebody walks through the door, I think that there is a synchronization between my brain waves and those from the guy that's coming in," he says. "Maybe it's that we're sharing the room or maybe it's that we share the same space at the same time. People with mindsets that are starting to synchronize with each other. That is the beginning of empathy."

The question is, can computer scientists like Ikegami create a machine that can get into synchrony with a human? That's what he's working on. His aim is to create an android that can pass what's called the Turing Test. The test was devised by British mathematician and cryptanalyst Alan Turing. His work during the Second World War helped the Allies decipher intercepted coded messages from the Nazis.

The Turing Test consists of a five-minute keyboard conversation between a human and a computer. For a machine to pass the Turing Test, people conversing with the computer or android must mistake the machine for a human being more than 30 percent of the time.

In 2014, a computer program named Eugene Goostman was one of the first to pass the Turing Test at an event organized by the University of Reading. Eugene Goostman, a chatbot created by Vladimir Veselov and Eugene Demchenko, simulated a 13-year-old teen living in Ukraine. In 2016, scientists at MIT were also able to beat the Turing Test.

ERICA doesn't pass the Turing Test, but that's only because her programming doesn't allow her to go deep into conversations just yet. But that's coming. Even so, I was blown away by how much I cared about what was going on inside her mechanical head.

Ikegami says the Turing Test (or something like it) can be used to tell whether a robot mind approaches that of a human. But here's the thing. Ikegami does not believe that the key to making robots empathetic or capable of forming relationships with people lies in making them intelligent. He thinks the key is to make them seem alive by making them both autonomous and unpredictable.

Kiyoko Kitagawa looks after old people in their declining years. Shigeki Sugano uses his experience with ailing parents to design robots that learn by observing things from an old person's point of view. ERICA's creator Dylan Glas spends more time thinking about empathy than almost anyone I've ever met. Hiroshi Ishiguro and Takashi Ikegami are pointing the way toward machines that fill the void created by loneliness.

I went to Japan in search of empathic robots and androids. What I found were kind people. I found humans with an astonishing capacity to care about robots and androids. I found *sonzai-kan* inside machines. And I began to rediscover it in myself.

CHAPTER NINE

Teach Me

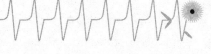

"Class, take your seats on the floor as soon as you can," says the grade 4 teacher in a commanding voice that is firm but friendly. "Don't forget that we have science period immediately following recess."

Twenty-three girls and boys file into the classroom and take their seats in a makeshift circle on the floor at the front.

"I'm going to turn things over to Ms. MacDonnell," says the teacher.

"Good afternoon," Angela MacDonnell greets the class. She sits on a chair at 11 o'clock on the circle. To her right are two empty chairs that are about to be filled.

"How many of you are looking forward to seeing baby Liam?" MacDonnell asks the students. All but two raise their hands.

"What are you excited to know?" MacDonnell asks.

"I want to know how much he's grown," says a girl.

"I want to know if he can stand up," says a boy.

"Well, let's find out," says MacDonnell, who walks over to the classroom door and opens it.

Liam is brought in by his parents, Tania and Michael. The boy who asked if Liam can stand up is going to be disappointed; the baby is two days shy of five months old.

"Let's sing a greeting for Liam," MacDonnell says as she invites the students to stand.

"Hello, baby Liam, and how are you, how are you, how are you? Hello, baby Liam, and how are you, how are you today?" MacDonnell starts singing a welcome song, and the children join in immediately.

On cue, Tania and Michael carry Liam around the circle so that each child can greet the baby with a personal hello—something each does with a smile and a surprising amount of focus.

Singing over, Tania puts Liam on the floor.

"What do you notice about Liam?" MacDonnell asks the class.

"His hair is growing in," says one student.

"His legs are getting a little bit stronger," says another.

Liam starts looking around.

"Do you think he needs a toy?" asks MacDonnell.

"Maybe he's reaching for his shoes," offers one of the children.

Liam beats the class to it by grabbing a nearby plush toy in the shape of a clownfish. He puts it in his mouth.

"Why do you think he did that?" MacDonnell asks.

"That's how babies know," says one of the pupils.

"Babies use their mouths to check things out," says MacDonnell.

I notice that the group leader says very little to the children; she lets them discover Liam's behaviour for themselves.

"Let's see if he likes this." Tania, Liam's mom, picks out a red, blue, and yellow soft toy in the shape of a power drill and puts it on the carpet a metre away.

Liam uses his chubby forearms to push his torso off the

carpet. He raises his right arm to reach for the plush drill, but the attempted manoeuvre makes him lose his balance. His chest flops to the carpet and he begins to cry.

"What's happening with Liam?" MacDonnell asks the students.

"He's frustrated," says one of the students.

"Maybe he's just fussing," says another.

MacDonnell doesn't judge the responses of the students. She just keeps the conversation going. "What things might make a baby cry?" she asks.

"A dirty diaper," says one student.

"He's lonely and he wants to be held," says another.

"What would you like to ask Liam's parents?" MacDonnell asks.

"Are there times when Liam won't stop crying?" asks a child.

"That happened last night," Liam's dad, Michael, replies. "His diaper wasn't dirty, and he wouldn't take a bottle. I tried talking to him."

"There must be times when the baby won't stop crying no matter what you do," says MacDonnell.

"That's when Michael pulls out his recorder," says Tania.

Michael pulls out a recorder from Liam's diaper bag and starts to play "Are You Sleeping?" ("Frère Jacques").

Liam snaps to attention, turns toward his dad, and smiles.

"Not every parent would try something like that," says MacDonnell.

And not every class gets a lesson in empathy like this.

"The male nurturing lessons Michael gave the students are amazing," says Mary Gordon, a former kindergarten teacher who is the creator of the program I just witnessed. "This will have a long-term impact on the children."

Long-term and life-altering impacts—these are the goals Gordon wants to achieve. The program is called Roots of Empathy (ROE). Gordon founded it in Toronto in 1996. Roots of Empathy aims to reduce aggression and bullying among schoolchildren by increasing empathy. The educator is trying to address a huge social problem. In a 2003 article published in the *Canadian Journal of Psychiatry*, Wendy Craig, professor and head of the department of psychology at Queen's University, and Debra Pepler, professor of psychology at York University, wrote that 18 percent of children reported being bullied in the five days preceding a survey.

Roots of Empathy can be taught in an age-appropriate way at any point from kindergarten to grade 8. Its ambitious curriculum is divided into nine chapters and 27 sessions, each focused on getting students to see things from a developing baby's point of view. Angela MacDonnell, the group leader whose session I observed in the grade 4 class, encourages the students to develop cognitive empathy by observing and identifying baby Liam's feelings and by trying to figure out what the baby wants and needs. The topic of the session I observed is "why babies cry."

Later, Mary Gordon invites me to her office in Don Mills to explain the guiding philosophy behind the program. Fittingly, the location is just minutes from the Ontario Science Centre, a museum that is all about engaging young minds.

"Cognitive empathy is perspective taking," Gordon explains. "We work really hard on perspective taking. For instance, we get down on the floor and ask the children what the baby can see. We want them to understand that from that position, the baby can only see your running shoes."

As Gordon explains it, that's just the start.

"From a physical way of helping the children understand perspective taking, you can get to cognitive perspective taking," she explains. "We get them to read the baby's cues. Do you think the baby is trying to get that toy? How do you know? What's the baby doing with his body? What's the baby's expression? What sounds is the baby making? Is the baby grunting? Things like that help them understand the baby's intentions."

The next step is priming the students to identify the baby's feelings.

"When we sing, does the baby look overwhelmed, excited, or a little bit shy?" Gordon asks rhetorically. "We ask the students what they think, so they learn to read the baby's cues."

Choosing a baby as teacher is the genius of the program. It enables kids who didn't experience or learn empathy in their own homes to travel back in time and witness its infant beginnings, before the baby has learned to talk.

"If you haven't had empathy through experience, I don't think we can fault people for not having it," says Gordon. "That's why I have Roots of Empathy: to give kids a chance to develop it if they didn't get it."

Babies like Liam are at the heart of the program; but so are parents like Tania and Michael. As Gordon explains it, putting babies and parents together allows students to witness a baby forming a loving attachment to his or her parents.

"All we are looking for is a secure attachment; not a PhD, not socioeconomic standing, not language but someone who lives in the neighbourhood and who could be a rock star because of how they love," she says.

How they love can make or break a child's capacity for kindness. Gordon knows all about that. As a teacher in Ontario in the 1970s, she worked at inner-city schools filled with families living in poverty, with many instances of domestic violence, teen pregnancy, and drug use. As a teacher, she saw illiteracy as a core problem and wanted to fix it.

That was what propelled her first big success at social entrepreneurship. In 1981, Gordon created Canada's first school-based parenting and family literacy centre. Free of charge, the program helps preschool children learn to read and make a successful transition to school. It also helps parents create a home environment that fosters the child's learning and development. The program was so successful that in 2007 the Ontario government adopted it as a best practice. At present, there are 172 such centres across 19 school boards throughout the province.

These places serve vulnerable families, immigrant and refugee families, and solo parents. Talk to Mary Gordon for a while and it's clear she has a special affinity for parents who have abused and neglected their children, just as they were treated by their parents. A gifted observer, Gordon began to see patterns common to families at risk.

"My a-ha moment working with a lot of domestic violence and child abuse and neglect and even sexual abuse was that the perpetrators were never monsters, but they all lacked empathy," she explains. "My sense was that the common denominator in cruelty is the absence of empathy. It might be so in genocides, or in any kind of gratuitous public violence or intimate violence, that there was an absence of empathy."

Gordon believes that a child who doesn't learn empathy at home grows into an adult who repeats the cycle.

"I work a lot with women whose children were abused by their boyfriends," she says. "They would pretend they didn't know because they were so desperate for attention or recognition or to be seen by the boyfriend that they would sacrifice their children. They just didn't have the capacity to understand their children's emotional or even physical suffering."

The big idea of using ROE to demonstrate a secure attachment is to break the generational cycle of violence and poor parenting. "I believe that we are all predisposed inherently to empathy, born with the capacity for empathy, but it blooms or fades in that attachment relationship," she says.

Gordon's core belief is both the basis of Roots of Empathy and an affirmation of the "nurture" part of kindness. That is hardly surprising given that it comes from one of the most nurturing women I've met in years.

The story of how she got that way begins about 70 years ago in the rough-and-tumble streets of St. John's, Newfoundland.

Mary Gordon was born in St. John's in 1947, one of five siblings. Gordon's mother was an artist and her father was deputy minister of labour in the provincial government.

"For Newfoundland, it was not a big family," she says. "But there was always a lot going on. People would be coming and going, and it was a very busy Grand Central Station type of house. Nobody was perfect. The children fought, and it was a totally normal house."

Gordon's parents gave her many gifts that set her on her professional path. Two stand out. The first was summers in Cherry Lane in Manuels, a picturesque place near Conception Bay South, about an hour's drive from St. John's.

"I think it was not so much solitude as freedom from having to be anywhere or do anything or act in any particular way," Gordon recalls. "And it was timeless. It was complete trust by all the adults that you would take care of yourself. We used to build rafts with nails and hammers, and nobody thought that was unusual. There were horses on the beach. We did things I would never let my own children do. There was a sense of care-free abandon. Mealtimes were wonderful. We had sunsets on the deck, and the family was together.

"I remember feeling free like it was the beginning of time, rather than any particular epoch of history," she adds. "And that I think was the deepest, happiest time frame I can remember."

The way Mary Gordon describes it, those summers sound a lot like the bliss that Theodore Fontaine experienced before being taken to an Indian residential school. In his case, one can argue that the early childhood bliss that he felt with his parents and extended family fortified his resilience to survive 12 years of physical, emotional, and sexual abuse. In Gordon's case, that "happiest time frame" provided a foundational counterpoint for the next gift her parents gave her—a lesson about her place in the world compared to people less fortunate.

"My mother was incredibly empathic in a very quiet way," says Gordon. "She was an artist, though she put her art on hold for the children. She did practical things that gave me insights into our shared humanity with people who were less fortun-ate. We would visit homes, and I would help her carry food and clothing."

The creator of ROE says that her mother would accept gra-ciously and without fail the hospitality of the people she helped. "My mother would sit down in the squalour and accept the hos-pitality of the woman with teeming children who ran around in a house with a dirt floor and no basement," she recalls. "Sometimes, they didn't even have running water. They had to go up the hill to the Battery in St. John's to get it. The Battery is the most beautiful place, with the most abject poverty."

One such visit stands out in Gordon's mind for the stinging rebuke her mother gave her after she was offered tea in a cup that appeared unwashed. "I remember turning the cup away to have my tea because the cups were very stained," Gordon recalls. "In the car afterwards, my mother said to me, 'What makes you think your germs are any better?'

"I'm just a step away from being in their circumstances," says Gordon. "I just happen to have landed lucky, and it was my job to understand other people's issues and to not think that I'm any better."

That guiding philosophy extended to people who had just gotten out of prison. "They would come to our house for a meal," recalls Gordon. "I remember many times being at the card table in the hall with someone who had gotten out of jail and who didn't smell too good. My mom would call that person our guest and tell me to sit down and keep the guest company."

Gordon says her father had a similar influence on her when she was growing up. "People in the hospital or in old age homes didn't get visited very much," she explains. "In Newfoundland, your family couldn't come in to St. John's and stay with you. Every Sunday, my dad would visit the sick and the dying. To spend time with him, I would often go with him on these visits."

She says her father would approach each patient in the hospital or nursing home and offer them the choice of a shared prayer with him or a song from her. "They'd always vote for the little one to sing them a song," she recalls.

On the drive home, she and her dad would talk about the people they visited. "He would explain that the person we'd visited was somebody's father, somebody's son, or somebody's uncle," she recalls. "He got me to imagine that the man might be lonely and perhaps worried about what was wrong with his health, worried about what was going on at home. He might be worried about not providing a livelihood."

Gordon's parents were giving her lots of practice at perspective taking. Also embedded in her parents' teaching was affective concern: a call to action to help less fortunate individuals close to home and (later) abroad. "They wanted us to be global thinkers in a time when there weren't global thinkers," she says. "My dad would get out the globe and show us where charitable funds were going. He would tell us they might need it because they don't have potable water or because they need it to send someone to school."

Gordon says her parents gained their insight through different paths.

"My dad had adversity," she says. "His brother Gwynn died an agonizing death in the bedroom next to him, screaming in pain every night. That scarred my father incredibly, and his wasn't such a loving family."

As a young man her father had his own brush with death. "He got meningitis and nearly died when he was 20 years old," says Gordon. "The sisters ran the Mercy Hospital where he was admitted. He was delirious but he could hear everything; as a doctor, you would appreciate that. He played the Messiah in a play, and the nuns used to come and visit him and say, 'He is such a nice young man. We'll come and pray for the Messiah.' I think near-dying affected him. I think the loneliness and isolation of childhood affected him, as did the suffering of his brother."

Gordon's mother too faced hardship starting at the age of seven, when her father drowned at sea. Her mother's extended family gave them emotional and financial help. To help make ends meet, Gordon's maternal grandmother took in soldiers during the First World War. She also ran a cafeteria across the street from a mental health facility.

"Every so often someone made a run for it because they would be taken on day outings," says Gordon. "They'd end up at our place. My grandmother would be very sweet to them and they would come over, though she slept with a bat just to protect herself."

Gordon says her mother learned something more than being sweet. "My grandmother taught her to help them because they were sick, and not to judge them. I think the absence of judging is a very big thing in the recognition of the dignity of people."

Gordon's mother found that financial hardship made her sensitive to the plight of disadvantaged people; living amidst a warm and loving extended family compelled her to extend kindness to strangers. Her father developed deep empathy for others who suffer from the pain of disease and loneliness because both were part of his life story.

It's striking to Gordon that her parents influenced her in different ways. It's something she's been pondering for decades. "You're getting at the roots of how we become who we are," says Gordon.

That of course is one of the main reasons why I embarked on a journey to learn the roots of empathy from people adept at it.

"I think we become who we are at home," she says. "Adults are always looking for that feeling of being at home. I think you are really shaped at home. A good experience there can buffer negative experiences in life. But if you don't have that good basis at home, you better have good teachers, and you better have empathic people in your life. Children who aren't seen and felt at home don't learn to see or feel themselves."

Gordon wants to change that. And it's not hard to imagine how those early childhood lessons from her parents have made the Roots of Empathy creator uniquely qualified to do it.

* * *

The ROE program is offered across Canada and in close to a dozen other countries. Mary Gordon says more than a million children have taken the training.

Does ROE make kids more empathic? The answer is maybe. The program was tested in Manitoba, and although the research data is limited the results so far are encouraging.

Compared to students who didn't get ROE training, those who did had significantly lower scores in physical aggression, things like kicking or hitting other children, threatening others, and overt bullying. They also had lower rates of indirect aggression, which included trying to get others to shun someone or telling a student's secrets to another person. Students who took ROE also showed higher rates of offering to help other children having difficulty, comforting a child who is upset or crying, and inviting shy outsiders to join a game.

The authors concluded that broad implementation of ROE could reduce fighting between children at school from baseline rate of 15 percent to 8 percent. The program was also found to be cost-effective. At the time, ROE cost $4 per child per session; that works out to $108 per child per year. The cost to the education system of dealing with a child with a conduct disorder is estimated at $7,944 per year, between the ages of 10 and 28.

The British version of ROE rolled out in 2010 as a partnership with the charity Action for Children, which focuses on vulnerable and neglected children. A study by researchers at North Lanarkshire Council's psychological service found that ROE significantly increased empathy and reduced aggression in Scottish students.

There is some evidence that ROE has long-term benefits. But

it's hard to get that data, since provincial privacy laws prohibit Mary Gordon from finding out what happens to kids who take the course. (Those rules are what compelled me to change the names of the ROE family I observed and to omit the name of the school that I visited.)

Sometimes, though, Gordon receives follow-up information by happenstance. That happened recently when she attended what ROE calls a Baby Celebration, an annual local event in which ROE parents and their babies are thanked for participating in the program. There, Gordon met a young man and his girlfriend who had volunteered as ROE parents. The man "remembered he had had [the program] in grade school," says Gordon. "His girlfriend said that she thought he was a better father because of what he remembered about Roots of Empathy."

As well, one of Gordon's ROE instructors was invited to speak at a teachers college in Atlantic Canada. One of the students at the session told his professor that he had attended ROE as a 10-year-old. The student said that a discussion he had one day at ROE made him realize that he was a bully. The night following the ROE session, the 10-year-old boy cried for most of the night. He decided then and there that he would neither bully nor be mean to anyone else again.

"He came back to school the next day and he didn't say much of anything," says Gordon. "But he knew that several people had a better day and a better year and a better life because he wasn't being mean to them. And he decided that he would be a teacher and that no one would be bullied in his class."

How sure is Mary Gordon that ROE made the difference?

"We know what changed his life," she says.

* * *

Embedded in ROE and thousands of other courses is the assumption that empathy and kindness are skills that can be taught.

It's not hard to see why the customer service industry tries to teach empathy to employees. A 2014 survey by American Express found that 60 percent of Americans per year walk away from an intended purchase due to a bad experience with customer service, up 5 percent since 2012. The same survey found that 37 percent of customers switch companies after one poor service encounter.

Tech giant Apple apparently puts a lot of effort into teaching kindness. In a 2012 article on Gizmodo.com, reporter Sam Biddle revealed excerpts he says were taken from an Apple training manual titled *Genius Training Student Workbook*. Biddle said the manual contained "Apple Dos and Don'ts, down to specific words you're not allowed to use, and lessons on how to identify and capitalize on human emotions."

As a watcher of empathy, I have no doubt that Apple employees have empathy imprinted on the brain. A couple of years ago, I bought an iPad and had to return to the store almost immediately when I couldn't get it to work. "That really sucks," said the first Apple employee I spoke to just inside the front door.

Biddle says that the word *empathy* is repeated "ad nauseam in the *Genius* manual." It is also filled with suggested lines that service employees are encouraged to use with customers. But are canned responses empathy?

"You can't script empathy," Bruce Temkin told Fortune.com in 2016. Temkin is a customer service consultant based in Boston. "The right way to do it is to teach the agents about why you're trying to show empathy, what is it, and why is it important."

Empathy courses are also gaining popularity in health care.

Some use videos of interactions between health professionals and patients. Others feature role play. Many focus on communication skills. One course teaches health professionals how to interpret the patient's body language. Some focus almost exclusively on training doctors how to respond to the concerns of patients.

The doctors who teach medical students are divided on the benefits. In an article published in 2016 in the *Journal of the Royal College of Physicians of Edinburgh*, David Jeffrey concluded that health professionals can be taught cognitive empathy, as well as how to share feelings and support one another. Another author, however, disagreed. "My own conclusion is that empathy cannot be taught," wrote Robert Downie in the same edition of the same journal. "It is undesirable to attempt to teach it, as a concentration on emotions can mislead, blind judgment and lead to bad outcomes."

You may be surprised that Mary Gordon counts herself among the skeptics.

"I don't believe you can teach empathy," she says. "I think it's caught and not taught. I think it's experiential. You can provide water but you can't make a person drink. We provide the water and the encouragement, and maybe we can get children to reach for it."

However, there's a new approach that's worth keeping an eye on.

Grit Hein is a young and up-and-coming neuroscientist at the University of Würzburg's Clinic for Psychiatry, Psychosomatics and Psychotherapy in Germany. She formerly taught at the University of Zurich, where she and her research partners recruited 40 men, evenly divided among native Swiss and residents of Balkan descent, to be test subjects. She wanted to study empathy in a particular social context.

Immigrants from the former Yugoslavia make up one of the largest ethnic communities in Switzerland. Residents of Balkan descent have found it hard to integrate into Swiss society, and Swiss citizens have long viewed them as purveyors of violent crime and other social problems. Many people of Swiss descent view themselves as members of an "in-group" and their Balkan countrymen as members of an "out-group." That dynamic provided the perfect backdrop to Hein's experiment.

In the initial part of the study, the test subjects were made to watch another person appear to receive a painful electric shock. Hein told the test subjects the identity of the person receiving the shock: Swiss or Balkan. While this was going on, the test subjects had their brains scanned with fMRI.

As they watched someone appear to receive a shock, each test subject's brain scan lit up in the anterior insula, meaning that the subject empathized with the shock victim. However, the Swiss nationals showed much greater empathy when a Swiss national appeared to get a shock and less empathy when the victim was of Balkan origin, a predictable result given the fact that humans are hard-wired to empathize more with members of an in-group than with members of an out-group.

But Hein wanted to see if teaching the test subjects to *think differently* about Balkans could change the results. In the next phase of the experiment, the test subjects were led to expect that they would receive the same electric shock that they had observed the victims receive previously. Now, however, each test subject was told that another person could spare him from the electric shock but that this person would have to sacrifice some of the money he had been given at the outset of the experiment to do so.

Just before the test subject was about to receive the shock— or be spared—the researchers gave the test subject the ethnic origin of the would-be rescuer, Swiss or Balkan. As before, the test subjects had their brains scanned to measure activity in empathy parts of the brain.

We expect members of our own ethnic group to help us out. So it should come as no surprise that when test subjects learned that a Swiss national of their own ethic origin had sacrificed financially to save them from an electric shock, their brain scans did not show more empathy toward them.

On the other hand, when the test subjects found out that a Balkan had prevented the electric shock, their empathy for Balkans, as shown on the brain scan, shot up. That means the test subjects, Swiss nationals all, had developed empathy for Balkan people.

At the end of the experiment, the test subjects said they liked Balkans about as much as they liked their Swiss countrymen. How fast did they change their minds about Balkans? It took just one or two saves from an electric shock!

"Based on everyday life experience we were surprised about the efficiency of our learning intervention," Hein tells me in an interview. "It was inspiring to see that it takes so few positive experiences to reduce lack of empathy toward out-groups."

You may be wondering what an experiment like this has to do with *teaching* empathy. This isn't didactic teaching, with lectures and seminars. Hein's experiment involves teaching that comes from life experience.

Learning theorists tell us that you're more likely to empathize with members of another group if you discover reasons to like members of the other group. As Hein writes in her paper, the fastest and most efficient way to make that happen is if a member of the out-group does something unexpectedly kind—for instance, a Balkan sparing a Swiss national an electric shock!

A surprising gesture of help, coupled with low expectations, is what drives people to change their opinion of a stranger. Simply put, when people you expect to be mean to you are kind, your brain is programmed to like them.

Like the participants in Hein's study, I've certainly warmed to someone on the basis of their unexpected kindness. Hein believes this type of learning in a lab has profound implications in the real world. "Given the current refugee crisis in Europe, one could test these findings in areas surrounding newly established refugee camps," Hein tells me. "For many people, the prospect of having a refugee camp in their neighbourhood is a source of anxiety. Maybe these anxieties could be reduced by 'staging' positive experiences for the respective out-group."

How you do that in the real world without making people feel manipulated is not an easy question to answer.

Hein has done a recent study showing that the same approach that increased the esteem with which Swiss nationals viewed Balkans can do likewise for the way patients view doctors and nurses. In other words, it could improve relations between patients and health professionals.

Could Hein's approach also foster better relations between doctors and nurses? The neuroscientist says the answer depends on whether a hospital pecking order exists. "The situation with physicians and nurses might be a bit more difficult, because here we are dealing with a clear hierarchy and differences in status," she says. "At least in Europe, nurses are expected to help physicians, and the support they provide is taken for granted. Thus, the help a physician receives from a nurse is unlikely to increase empathy toward the nurse."

If Hein is right, an important step in boosting empathy in hospitals is to get rid of power imbalances between physicians and other health care providers, patients, and family members. Some hospitals have chucked the hierarchy in favour of teams of health professionals of equal status. They seem to have more empathy. Still, in many hospital cultures, "team-talk" is little more than lip service; without a change in culture, the hierarchy and the lack of empathy remain.

In hospitals, thankfully, we neither give nor receive electric shocks. However, we do judge, shame, embarrass, and humiliate one another quite frequently. Sparing someone that sort of treatment might just be surprising enough to boost kindness among the people who toil in hospitals. "Pain is not necessary," says Hein. "In principle, any kind of positive unexpected experience with an out-group should work."

Here's an idea. Why not seek out someone you can't stand and do something for them that is both kind and unexpected? Right now, without waiting. There's no guarantee the other person will reciprocate. Who cares if they don't? You're unleashing some goodwill on the world.

The other life lesson to take from Hein's work is to set ourselves up for kind surprises by lowering our expectations of others.

When I started writing this book, I was dismayed whenever top experts declined to be interviewed. I wondered if they were trying to tell me that the book wasn't worth their time. I had such despair that at one point I considered giving up.

Two things turned my outlook around. First, I broadened my search for experts to interview. Second, I started empathizing with the people whose precious time I was asking to take up. Some agreed to speak with me. Others said no but were a bit kinder about it. The research really took off when I stopped expecting experts to say yes and when I was demonstrably grateful for any time they had to spare.

If you want to be blown away by the kindness of others, expect nothing. And if you want to blow others away by your compassion, show kindness to someone who you think least expects it from you.

Back at the grade 4 class, the Roots of Empathy session is almost over. Baby Liam looks tired, but the class's attention is still riveted on him.

"Class, let's everybody stand up and sing a goodbye to baby Liam," says ROE instructor Angela MacDonnell.

Liam's dad, Michael, picks him up and gently hands him to mom Tania. The threesome walk around the circle, stopping briefly to receive goodbyes from each pupil.

"Goodbye, baby Liam, see you soon, see you soon, see you soon. Goodbye, baby Liam, see you soon, see you very soon," sings the class in unison. They break into spontaneous applause as Liam and his parents complete the circle.

Mary Gordon has been watching intently. She is impressed that the students sat through the entire session without needing to get up and walk around.

Half the students cluster around Liam and his parents. This is an unscripted moment for them to observe how Liam interacts with his mother, Tania.

"Liam put my hand in his mouth," says a young girl.

"He does that a lot," says Tania. "Soon it will hurt when his teeth start coming in."

The remaining students group around Roots of Empathy creator Mary Gordon. "What have you learned from your time with baby Liam?" she asks.

"Ms. MacDonnell taught us never to shake a baby," says a young boy.

"Thank you," says Gordon. "How do you know that Tania and Michael are Liam's parents—and that they belong together?"

"You could ask someone in the school office," says a girl.

"You know by looking at them," answers another student.

"You'd be lucky to have a mum and a dad," answers a third child.

It's a quiet but telling admission from one of the students. Mary Gordon is hardly surprised. I find myself thinking that there must be kids who come from broken homes who see the secure attachment between a baby like Liam and his parents and feel sadness or envy.

"In every classroom, there are many children who have not had the benefit of a happy home life," says Gordon. "We know that there are children who are in foster care. We know that there are children who are living in abusive homes. Children are no fools. When they see bliss in this loving relationship between the baby and the parent, they know if they don't have it. We have had children say, 'My mom doesn't love me.'"

Gordon is thinking of a child who attended ROE a couple of years ago.

"After one of the classes, the child followed the mother out of the room," she recalls. "The mother asked if the little boy wanted to tell her something. That little boy asked the mother where love comes from." She shakes her head as she tells the story. "He's seven years of age and he knows he doesn't have it. I think we underestimate children's capacity for emotional understanding."

Gordon says her program can't fix broken homes. Still, she believes that the experience of being in a Roots of Empathy class for 27 sessions over an academic year gives children at risk an opportunity to observe up close a more positive kind of home life.

"It's a whole year of deep, reflective, interactive experience; they've laid down tracks in their brain of what love looks like," says Gordon.

For some reason, Gordon's words stick in my head as I drive away from the school. I don't come from what you might call a broken home. My parents fed me, clothed me, put a roof over my head, and sent me to school. But where other homes had hugs and laughter, I remember the loud arguments and frequent sighs that only much later I would come to understand as unhappiness.

Something about seeing Liam being held by Tania and Michael left a gnawing disquiet within me—as if I had missed an important foundational lesson about kindness. Suddenly, I remember watching TV shows as a child and feeling a pang of envy when the show depicted a happy family.

Mary Gordon's Roots of Empathy may not teach empathy. But it's still powerful.

CHAPTER TEN

Soul Whisperers

A plume of suffocating, sticky air envelops me the moment I exit the sliding doors of the Holiday Inn. I'm in Leader Heights, a sleepy community in a large agricultural region in south-central Pennsylvania.

Leader Heights is named after the late George Michael Leader, Democratic governor of Pennsylvania from 1955 to 1959. Following a career in politics, Leader switched into senior care. He founded Country Meadows and Providence Place retirement communities.

I step over the curb bordering the rear parking lot and walk down a steep grassy knoll. In front of me, I see a complex of three flat-roofed buildings in a rough V-shape that are set nicely into the valley. This is Country Meadows of York-South, one of 11 branded retirement communities located in Pennsylvania. I'm here because I've been told they have a way with people who have dementia, a very empathic way.

In the lobby a woman comes forward to meet me. "Hi, I'm Mandy Knight, associate executive director here at Country Meadows." Knight is also in charge of the facility's memory support program. She takes me to the second-floor residence called Connections. It's a long hallway with a central area for personal care assistants, nurses, and other health care personnel; a library; a meeting room for programs; and a kitchen for residents and their families to practise cooking. Each resident's room has a miniature version of a house front door; some are decked out with American flags, some with plush stars.

"We have about 41 residents right now," says Knight. "And that's full. There's a wait list that fluctuates."

The place looks warm and friendly. The floor is brimming with residents and staff, but green carpeting, acoustic ceiling tiles, and walls that absorb sound make it calm and quiet.

Currently, more than 5 million Americans have dementia, and Alzheimer's disease is the most common form. By 2050, the number of cases could rise to 16 million, according to the Alzheimer's Association.

My mother was diagnosed with dementia when she was 70. Like many people diagnosed with Alzheimer's, she ended up living in a nursing home. To the people who work in long-term care, those with dementia act in stereotypic and repetitive ways that seem baffling. An agitated woman yells the same phrase over and over again. A man says repeatedly, "I need to leave. I'm late," only to be reminded repeatedly that he isn't late because he has no pressing place to go. Here at Country Meadows, I'm hearing none of that. Each resident I see has dementia yet is calm.

Out of the corner of my eye, I see a young member of the staff, named Lorrie Quick, who is on her knees so she can be at eye level with a frail-looking elderly woman. The woman, who has dementia, wears a navy fleece jacket and khaki slacks. She sits in a wheelchair. The two are lost in conversation.

"I'm ready for a nap, Dorothy," says Quick. "Will you take one with me?"

"No, I think I already had one," Dorothy replies.

"No way," Quick protests with a look of mock seriousness. "You napped without me?"

"Take a nap now," the older woman replies.

"I think it would be a good idea," says Quick.

"I would if I didn't have to go back there," says Dorothy, pointing in the general direction of her room down the hall.

"That would be a lot of work," the care worker agrees.

"You can take a rest now," says Dorothy.

"Are you gonna take a rest with me?"

"Well, I can't take a nap in this sit-up chair."

"Is it my turn for the sit-up chair?" Quick counters.

"You don't want to get out of here so you can sit awhile," says Dorothy.

"I don't know what I'd do without you," says Quick. "You're a good mom."

"I practised," says Dorothy. "I learned how to be a good mom from all these girls that run around here."

"Are they all good girls?" Quick asks.

"Yeah," says Dorothy. "They all know how to rest."

"They do?" Quick asks. "Is that important?"

"If you don't rest, you fall over," says Dorothy, satisfied by her point.

"I love you, Ma," Quick replies.

"That'll do it," says Dorothy.

The exchange between Lorrie Quick and Dorothy reads like a scene from Samuel Beckett's absurdist play *Waiting for Godot*. It sounds random, but it's not.

In my experience as an ER physician and as a son, conversations between health professionals and people with dementia tend to be tense and argumentative. The person with dementia says something that sounds absurd, and the health professional corrects or contradicts, often impatiently.

I detected no such conflict between Quick and Dorothy; in large part Quick was deliberately following the things Dorothy said even if they seemed absurd, without judgment. The two seemed to be enjoying the moment together.

Dorothy seemed placid and friendly. But what if she isn't? Dementia patients sometimes have bouts of agitation and anger. "We have something called mirroring that we do, and you need to know your resident well enough to know whether you can do it or not," Quick says.

I first heard about mirroring from developmental psychologists; it is the hard-wired process by which mother and newborn baby bond to one another. It is also the way that human beings who like one another begin the process of becoming friends. It's intriguing that Quick talks about mirroring in the relationship between resident with dementia and care provider.

"If she's angry, you're angry too," she says. "If she gets mad and stomps her feet, you get mad and stomp your feet too. It might look ridiculous, but trust me, it's the fun part of the job. I can look like a complete idiot, but it does everything for them. They know that I understand—by reading their body language, their tone, and everything."

I'm struck that Quick knows what to do when Dorothy becomes sad or gets agitated. When my late father-in-law became angry, I wasn't sure what to do. When my late mother lashed out at my late father, he didn't know what to do either. It may surprise you that in most other places, nurses and personal support workers with years of experience don't know how to relate to people with dementia.

Health professionals get exasperated telling someone with dementia over and over again what day it is, or that their dead mother isn't coming to visit. And people with dementia get more and more frustrated, which creates even more tension for health professionals who don't know how to de-escalate the situation.

But Quick evidently does know. "Dorothy trusts me, she likes me, and she knows me," says Quick. "She might not know my name, but she knows me by my face and my actions and what I do and how I approach her."

It's the epitome of empathy for people with dementia, a kind of poetry in the moment between caregiver and old woman.

What matters is that Quick and Dorothy are in a synchronized *pas de deux*; the old woman with Alzheimer's disease leads and the young woman who cares for her follows. What they share is spontaneous but not accidental. It's part of a carefully crafted method for interacting with people with dementia, an approach that teaches more about relating to people with that dreadful affliction than anyone has taught me before.

The method is called Validation. It's a way to communicate with people who are disoriented, a sad but inevitable part of the later stages of dementia. In 1999, Country Meadows became the first organization to sponsor official certification training in Validation.

When the method works, Validation reduces stress and agitation. Its secret is the very essence of empathy, for success depends on developing an ability to see things from the chaotic and disoriented view of someone with Alzheimer's disease. Part technique and part philosophy, Validation is based on a disruptive premise: Instead of trying to bring the person with dementia back into our reality, it's better for them and for us as caregivers and family members if we enter theirs.

Validation is the creation of a remarkable woman named Naomi Feil. I call her the soul whisperer because Feil has an uncanny knack to peer deep inside the hearts and minds of people with dementia and see things no brain scan can possibly visualize. Her gifts include intuition, an ability to fuse the best of modern psychology with her own common sense and her boundless empathy for people with dementia.

Feil, who got her master's degree in social work from Columbia University in New York, began working on Validation in 1963. She was dissatisfied with traditional methods of working with people with dementia who are severely disoriented and agitated. Back then, frail seniors with Alzheimer's were often strapped for hours at a time in a Geri chair, a positional recliner on wheels. Nowadays, physical restraints are frowned upon.

"Today, of course, they would medicate and they would tranquilize the person," says Feil in an interview from Eugene, Oregon, where she lives when she isn't teaching Validation. "Tranquilizing medication is a barrier to Validation because if people have to swallow their feelings, then they're nothing. The other techniques that are often used—like redirection, diversion, and the therapeutic lie—also make people swallow their feelings."

By redirection, Feil means changing the subject. Diversion means offering up an activity or a glass of juice. The people who care for those with dementia use these two techniques when they can't answer the person's question or request and don't feel like lying about it. Then again, sometimes, they resort to the "therapeutic lie." Feil gives the example of a person with dementia repeatedly asking for her deceased mother. The therapeutic lie has the care provider assuring her that the mother is around the corner and will be right back. Feil was among the first to argue that such lying is not just unethical but pointless. "The old lady sits down because you're saying her mother is alive and she wants her mother to be alive and she sees her mother with her mind's eye," she says. "But on another level of awareness, deep down, that woman knows that her mother died because she was there at the funeral."

A 2013 study found that more than two-thirds of psychiatrists in northeast England had either sanctioned the use of lies by care providers or lied themselves. Care providers steeped in Validation are trained not to lie to residents. But they don't yank a resident reliving a time from the past back to the present either. Instead, they enter the resident's reality by asking questions such as "What do you miss the most about your mother?" or "What do you want to tell your mother?"

Questions such as these avoid lying or redirection, yet allow the person with dementia an opportunity to explore feelings and longings.

"If you listen and you really have empathy with the person at that moment, two minutes later that same old woman will say, 'My mother is with the dear Lord. She's not with us anymore,'" says Feil. "Once they've expressed their feelings, then they don't have to see their mother anymore."

Feil says the impossible-to-fulfill wants and desires of dementia patients are fleeting. Fight them and they get stronger. Acknowledge them and they dissipate in a few minutes.

Feil built a compelling hypothesis about what people with dementia want at the end of their lives: to return once more to the things that defined them. For some, it's a job, a profession, or a calling that made them proud. For others, it's a road not taken or a lifelong conflict with a significant family member.

"They're trying to resolve unfinished issues—mostly with their parents," she says. "Anger, feelings that they've never expressed, sexual feelings that have never come out but that need to come out in their old age."

And come out they do, in words or in gestures, in behaviours that vary depending on how advanced the dementia is. Without knowing the meaning behind the words and gestures, it's easy to assume they are made-up gibberish. Feil says they reveal what's on the mind of dementia sufferers: what it is that they're trying to relive, reframe, or resolve.

Along the way, Feil discovered that when words fail (and they almost always do), people with dementia use gestures to symbolize the feelings that they are trying to express. "One woman was always kissing her hand," she recalls. She deduced that the woman was expressing love. In a moment of inspiration, Feil pointed to the woman's hand and asked her if she loved it.

"The woman looked at me and said, 'Oh, my little baby. She's such a dear, dear girl.'" Feil recalls. "Her hand symbolized her baby. I found out that this particular woman had lots of children. She was almost 50 when she had her last child. She didn't pay enough attention to her daughter and she felt guilty, but she never told her when she was younger. She had kept that in, and now in her old age, she was using a part of her body to express sorrow so that she could die in peace."

Some people in the resolution phase of dementia seek to restore their identity. She recalls a man who kept pounding his fists. The man's son wondered why he did it.

"Your dad was a master carpenter," she recalls telling the son. "He doesn't see his fists—he sees a hammer and he sees a nail. Your dad is working. He's maintaining his identity. He's restoring who he is. He is never going to die [like] a living dead person in a wheelchair. He's going to die a master carpenter."

My father-in-law, Gabriel Broder, was diagnosed with Alzheimer's disease in his eighties. A self-made man, Gabe grew up in Montreal before moving to Winnipeg to start a business and raise a family with his wife, Phyllis. In the 1970s, he sold his successful fabric store and moved his family from Winnipeg to Israel to work as a farmer in a development town in the northern Galilee region of the country. Like a lot of other newcomers, Gabe found that doing business in Israel was a lot different from what he was used to in Winnipeg. In the 1980s, the family moved back to Winnipeg to start over. Gabe and Phyllis moved to Ontario nearly 20 years ago to be closer to their children and grandchildren.

Gabe began to show signs of Alzheimer's in 2010. In 2013, he moved into a long-term care facility located in northern Toronto. From that day forward, almost every time I visited Gabe, there'd be a moment when he would stand up and say, "Let's go." He would say it with great purpose. I never got a clear sense of what he was in a hurry to do. Neither did his wife, Phyllis, my partner, Tamara, or the rest of the family.

The fact that I couldn't crack the Enigma code of what made Gabe so anxious to go someplace important drove me crazy. Without knowing the intimate and sometimes painful parts of a person's story, it's next to impossible to tie an odd behaviour like banging fingers against the side of a wheelchair to an old and unresolved hurt or a conflict.

It's a challenge that Naomi Feil has relished for decades.

"You rarely find out what it is," says Feil. "Still, when someone is banging, it's not hard to notice and say, 'Boy, are you angry. Did they hit you?' If you're lucky, they reply, 'Goddamn sonofabitch. Yes, he hit me.' If he does, I ask, 'Did he hurt you?' If he nods or answers yes, I reply, 'What do you want to say to him?'"

Feil says it might be a father, a mother, a sibling, a cousin, or a teacher who hit the patient. It's important to recognize that the gesture is symbolic of something that has a powerful and emotional meaning. The Validation method relies on these intuitive leaps. It all stems from Naomi Feil's astonishing capacity to put herself in others' shoes. No brain scan can uncover these unresolved feelings.

What made Feil a soul whisperer to people with dementia is what I am determined to discover.

The short answer is that Naomi Feil grew up living in a nursing home. Not many people can make that claim.

She was born in Munich in 1932. Her parents, Julius and Helen Weil, decided to leave Nazi Germany four years later.

"My mother used to tell the story that a neighbour came and told her the Gestapo had our family on a list, and told us to leave as quickly as possible," Feil told the German-language magazine *Amerika Woche*. "My father was the director of a Jewish home for troubled boys. The Gestapo came and searched the home so my father knew it was time to go."

Julius Weil left Germany in 1936; Helen, Naomi, and her sister followed a year later. Julius, a practising psychologist, arrived in New York City and got a job counselling at-risk young women. In 1940, the family moved to Cleveland, where Julius got a job as executive director of the Montefiore Home for the

Aged. There he pioneered the hiring of occupational and physical therapists. Helen also worked at Montefiore, becoming one of the first social workers in America assigned to a nursing home.

In addition to working at Montefiore, Feil's parents resided there. With no separate living quarters for staff, they lived among the residents, all aged 65 and older. Here was Naomi Feil, barely eight years old, living right alongside people with dementia. At first glance, it seems utterly strange, maybe even a bit pathological. But the little girl's life story to that point in time was unusual. "We had just come to America from Germany, and I didn't have many friends in school," she tells me. A Jew and a German ex-pat, Feil says both made her the subject of taunts at school.

Feil was cut off from her familiar childhood places in Munich. In America, she felt the sting of prejudice from fellow students and found herself living in a strange new place with little if any emotional support from her parents, who were too busy running the nursing home. Her salvation, it turns out, came when she began to recognize that she had something in common with the nursing home residents: loneliness.

"I found the residents to be very warm," she says. Some of them invited her to their rooms. "One of the ladies taught me French. I would run home after school and I would get a French book, and Madame Ramensoff—who came from Alsace—taught me."

Young Naomi grew to love the place. Looking back, Feil realizes that the seeds of her empathy for old people began there. And the key was that she experienced Montefiore with the eyes of a child.

"As a child, you have a different intuition," she says. "You have a different gift for relating to people than you do as an adult or even as a teenager. I was a little girl with a need to have friends, to be loved and to give love. I think that had a whole lot to do with it. I have a natural empathy for very old people that I don't have for all the other ages."

Feil had many friends among the residents, and one special companion named Florence Lew. Mrs. Lew was her best friend at Montefiore. You have to picture it. At the time, Mrs. Lew was 68 and Feil was just eight. "She was tall and well built, with a fine, longish thin nose on which she perched her bifocal lorgnette," wrote Feil in the prologue to her book, *The Validation Breakthrough.*

Feil describes a pivotal meeting between herself and Mrs. Lew, one that cemented their friendship. The old woman found the eight-year-old girl with her feet tangled in roller skate straps, sobbing on the cracked pavement leading to Montefiore. The little girl told Mrs. Lew how hurt she was that her mother had given her sibling a better pair of roller skates.

"To heal my hurt, she produced her diary, which she always carried in her big, black, shiny purse," wrote Feil. Mrs. Lew found a passage from her diary and began to read it.

June 10, 1891

Dear Diary,
My mother hasn't changed. She embarrassed me again today, just like she did in Miss Nelson's third-grade classroom.

"She told me about her life, and how her mother hurt her," Feil recalls. "I told her how my mother hurt me, and we had a bond."

Their friendship grew, and the two became inseparable. "We'd take long walks," Feil recalls. "We went to the movie theatre about three miles each way. We got in free because we were from the Home. We saw Captain Marvel and all those movies. After, we got ice cream. We loved each other, actually."

That love included confiding in one another. Reading from Mrs. Lew's diary, Feil learned more about one of the older woman's deepest childhood hurts. As a young girl, she had a beloved toy rabbit made of wood. "She and her father had made it together," says Feil. "Her dad tied it to a string and gave it to her. She loved it. She called it Creaky because its paws creaked when she moved them."

In a pivotal diary entry, Florence Lew told the young Naomi Feil about the day her mother came to the school that Lew attended to take the wooden rabbit from her.

My mother grabbed Creaky so hard that his hind leg snapped off. She marched up front and threw Creaky into Miss Nelson's steel wastebasket. Creaky made a hollow sound as he hit the bottom. I ran up to save him. Miss Nelson took the wastebasket with Creaky away.

Feil asked Mrs. Lew what happened after her mother took Creaky. "I died," was Mrs. Lew's reply.

Over the next 12 years, Naomi Feil and Mrs. Lew were inseparable. Together, they sang songs, went to the movies, and picked rubber from tires for the U.S. war effort.

In between those tender moments of deep friendship, Feil became aware that Mrs. Lew was not just a resident of Montefiore but a patient. "Every once in a while, she would disappear," she recalls. "And that was when she went to get shock treatments. Everyone said she was crazy. I knew she wasn't crazy. She was just my friend."

In 1950, Feil, by then a young woman, travelled to New York City to study psychology and social work. After graduation, she worked in New York for seven years before returning to Montefiore in 1963, this time to work with residents in the late stages of dementia.

In the early 1940s, Julius Weil had started a special unit at Montefiore for residents with dementia who were disoriented and extremely agitated. It was there that Naomi Feil met her old friend Mrs. Lew for the first time in over 13 years. Feil was 31; Mrs. Lew was close to 91, if not older. "I didn't recognize her at all," says Feil. "Her eyes were blank. She was restrained in a Geri chair and she had no communication with the outside world at all." Mrs. Lew kept uttering the same nonsensical word over and over. "All she said was, 'Pree, Pree,'" Feil recalls. And that is the moment when Feil made an astonishing leap of intuition.

"I replied, 'Cree, Cree?' And then, she looked at me and she recognized me."

The memory of Mrs. Lew reading from her diary, and telling her about Creaky the wooden rabbit, and how hurt she felt when her mother took the rabbit away, all came flooding back. In that moment, Feil wasn't the objective therapist, but an old friend who was able to see things from the point of view of a frail elderly woman who was clinging to the painful memory of the day her mother had disposed of her beloved wooden rabbit Creaky. Feil also remembered how Mrs. Lew had acted during the happy times the two had spent together.

"She needed to walk," recalls Feil of Mrs. Lew. "She wanted me to get her out of the chair. I tried to help her do that, but the nurse wouldn't let me."

Mrs. Lew died the same night. After she died, Feil realized what the elderly woman was trying to say in those last days.

She deduced that in the final stages of her dementia and her life, Mrs. Lew was trying to express the anger she felt toward her mother, and the deep sense of loss she experienced when Creaky was taken away.

Following Mrs. Lew's death, it took another 17 years for Feil to put all of the pieces of the Validation approach together, but what she had figured out in the wake of Mrs. Lew's death was the main thing. "What she taught me was to listen and to respect the speech, the language, the movements, and her need," says Feil. "Her human need was to express first her anger at her mom, and her need for love—to understand that there's always a reason behind what people do."

That initial insight came in 1963. Today, Feil, herself an 85-year-old woman, feels deep regret that she wasn't able to help Mrs. Lew. I disagree respectfully. I believe that in recognizing the meaning behind her cry of 'Pree, Pree,' she validated Mrs. Lew's emotional pain and made it possible for the old woman to die in peace.

Slowly, Naomi Feil turned her prodigious capacity to empathize with frail seniors with dementia into a method to care for them by understanding them. In 1982, she published her first book, *Validation: The Feil Method.* Her second book, *The Validation Breakthrough*, was published in 1993, and updated and revised in 2002.

By the 1980s, she was teaching the method to health care workers throughout North America and around the world. But how much of her method can be taught to others? Can I learn to be as empathic as Feil? Country Meadows is a telling place to find out, because in 1999 it adopted the Validation method.

Stephen Klotz, a former parish priest who doffed his robes back in 1995, is the executive director of Validation education for Country Meadows, chain-wide. Klotz says Validation was a natural fit for the retirement community. Naomi Feil was brought in to train the first 12 employees, who included Klotz. They were to be the guinea pigs who would learn the technique and then train the others. They got to see Naomi Feil in action.

"She'd go over to this woman and introduce herself and take her hands and look into the woman's face and say, 'You look sad,'" he recalls. "The woman might say, 'I am sad.' Naomi would say, 'Who do you miss?' And every time, the person would name someone. Oftentimes, Naomi's next question as she moved in even a little bit closer and maybe put an arm around her was, 'And what do you miss about that person the most?'

"*Boom!* In a matter of a few seconds, she was into that person's world." Klotz is reliving the moment as he speaks. "She had created this little bubble in a busy hallway. It was as if the rest of the world didn't exist. It was just Naomi and this older woman, and the older woman was opening up to her about this pain that was inside. That really convinced me of the truth of what Naomi was saying and that she does practise, and has practised, what she's preached."

It's 10 A.M., and eight residents are gathering for their weekly group session. They sit in a tight circle facing one another.

"All right. I'm going to ask us all to join," says Christina Lawrence, a personal care assistant and the coordinator of the group. "Maddie, will you welcome everyone to our group today and tell us to begin our meeting?"

"I want to welcome everyone to the meeting today and hope that we have a constructive meeting," says Maddie, a resident with dementia who acts as the group's designated leader.

"Very well said," Lawrence replies.

After a chorus of "You Are My Sunshine," the meeting gets to the heart of the matter.

"We've got a good group of ladies. Earlier, I gave Miss Maddie an entire book's worth of quotes to read over and this is the one she picked for us today. Maddie, will you read that aloud for our group?" the coordinator asks.

"'Friendship is not something you learn in school,'" Maddie reads in a full voice. "'But if you haven't learned the meaning of friendship you really haven't learned anything.' That's written by Muhammad Ali."

"Thank you, Maddie," says Lawrence. "What do you like about this quote? What stuck out to you?"

"Friendship is something that you learn as you get older," says Maddie.

"What's a way you can be a good friend?" Lawrence asks the group. "How can you be a good friend to someone?"

The vibe in the room seems friendly. Most of the group members seem content to sit back. A woman named Fran shifts in her seat. "Get to know people and how they are or what problems or, you know," she says. "Then, you start getting together."

When no one adds anything, Lawrence goes back to Maddie, the group's leader and most articulate resident. "What do you think, Maddie? How can we be good friends to people?"

"You have to be a good friend in sharing what you have learned and share that with everyone that you come in contact with," Maddie replies.

"Is it important to share when you're being a friend?" Lawrence asks. "Mary, what's a way to be a good friend? We mentioned learning about people and sharing."

"That's a lot of it, but you have other things too," Mary replies. "Things that maybe aren't out in the open and you know about them or anything in their nature."

Lawrence smiles, sensing that Mary is dialled in to the conversation. She asks Mary a follow-up question. "They can share stuff with you, right?"

"Right, and I think that's a good reason," says Mary.

"How would we feel if we didn't have our friends around?"

"Lonesome," says Mary.

"Lonesome," Lawrence repeats. "Pauleen, what would it be like if we didn't have our friends around?"

"It'd be awful," says Pauleen.

After some more discussion, refreshments, and a chorus of "Goodnight, Ladies," the group adjourns.

The content sounds banal until you appreciate that everyone in the group except Christina Lawrence has advanced dementia. She says this kind of group activity is Validation in action. "It's about being with them in the moment [with] what they're feeling," she explains. "If they start to go away from the direct question, you just go with it because that's their thought and that's what they're sharing. If someone says something a little off the wall, you don't correct them. You stay in the moment with them."

Lawrence did more than share the moment with members of the group. She made eye contact with each member. She took turns being warm with each of them. I also noticed that she touched some but not all of the women, in a gesture called anchor touching.

"You give your friend a pat on the back, it's warm and comforting," says Lawrence. "Doing those little touches evokes that feeling of a friend or a parent or whatever the touch corresponds to. For most people with dementia, it's comforting to have touch."

Each member of the group I watched has been selected for a specific role, and none is more critical than the group leader, Maddie. "She is our highest-functioning member," says Lawrence.

But even a high-functioning person with Alzheimer's disease may have quirks that affect her behaviour. In Naomi Feil's classification of the stages of Alzheimer's, a high-functioning person may be in a phase she calls "malorientation." From the standpoint of dementia, such a person might get angry and scold the group if it strays off the topic of discussion as established by the group leader and by Lawrence.

Maddie, says Lawrence, is not like that. "She's amazing. If I have a wavering moment, Maddie will ask, 'Why don't we do this today?' It's really neat to see that she's comfortable addressing the whole group and taking authority. It's the perfect role for her because she feels confident in it. It gives her purpose."

At Country Meadows, such judgments are not guesses but are based on knowing the resident.

"Maddie did secretarial and clerical paperwork," says Lawrence. "In her career, she was an organized person. Even to this day, she'll cut things out of the paper, and have things filed."

Stephen Klotz says the group is an essential part of Validation.

"All older people with dementia come to the point where they are validating each other, because oftentimes in a group, there's at least one or two people who are a little bit more attuned emotionally," he says. "Often, they will be the first to respond to a member who is sad, afraid, or feels another kind of difficult emotion. Then it becomes the group members and not the staff member who do the validating. Really, who can understand an

80- or 90-year-old with dementia better than another 80- or 90-year-old with dementia?"

I've worked on many geriatric wards, and I've visited many long-term care facilities. Until Stephen Klotz said it, I'd never heard any other health provider speak of people with dementia empathizing with one another.

You can't empathize with someone you don't know, least of all a person with dementia who has lost a lot of memories. Feil taught the people at Country Meadows to assemble a detailed dossier on the residents' life stories, their careers, their families, their hobbies, their likes and dislikes. Any small detail provides a possible clue to a behaviour that indicates that a resident is working through an emotional issue.

The heartbeats of Validation are the spontaneous one-on-one interactions between staff members and residents. Klotz says it took a lot of training for the method to click with him and his colleagues. As a former pastor, he was used to being the expert who points people in the right direction. With Validation, he had to learn to help people with dementia express their feelings without judging, correcting, or trying to influence them.

He remembers vividly the first time it all came together for him. "I was walking down the hallway with a resident named Ray," Klotz recalls. "As we were heading to a meeting room at the end of the hall, Ray asks me, 'Are we going to see my grandmother?'"

Klotz knew that Ray's grandmother was long deceased. "How often did you go to see your grandmother?" he asked. "Oh, we went to see her every week," the man replied.

As Klotz recalls, in that brief exchange, he noticed that Ray had switched from speaking in the present tense to speaking in the past tense. To Klotz, it was a big clue that even though Ray asked about seeing his grandmother, he also knew that she was dead.

"What did you enjoy about that?" Klotz asked the older man, also using the past tense, a subtle but important kind of mirroring that showed Ray that Klotz had followed him into the past.

"Her cooking and her baking," said Ray.

"What was the best thing that she made?" Klotz asked.

Ray stopped in his tracks and took a big sniff. There was no smell of baking in the air, but Ray imagined it.

"That was the big 'a-ha' for me," says Klotz. "Walking down that long hallway had transported Ray back to a memory of going to see his grandmother at the nursing home, and what a sweet experience that was. It was so pleasant that he likes to go back there sometimes."

Since then, he's seen lots of little moments. A woman who hadn't spoken for months suddenly speaking again after joining a Validation group; another woman who went from calling out constantly the name of her son to becoming the group's song leader.

He says that many of his colleagues have had major breakthroughs with their clients too. One noticed that a resident was holding and caressing her own upper arm. "Who is that?" the colleague asked the resident, inspired by Feil's story of the woman who repeatedly kissed her own hand.

"It's my little girl," said the resident.

"What happened to your little girl?"

"She died."

"How did it happen?" the worker asked.

The resident told the worker that for many years after she and her husband got married, they couldn't have a child. Finally, they had a baby girl who brought immense joy to their lives. At the time, the couple was so poor they could not afford a car. Finally, they saved up enough money. The woman's husband picked up the new car on his way home from work. As he pulled into the driveway, their daughter, now a toddler, was killed when she ran out of the house and into the path of the new car.

I start to cry as I hear Klotz tell this story. As a parent, I can't imagine the anguish that those two people felt.

With tears in her eyes, the resident told the worker that her husband had told her that after the little girl's funeral, they would never discuss it again. "The resident had carried this pain most of her life," Klotz recalls the worker telling him. "As she got older with dementia, she could no longer hold it in. Her grief started to seep out, and here was one of my colleagues validating her expression of sadness."

Klotz says that from that day forward, the woman was different. She continued to feel sad from time to time, but it was a sadness filled with meaning. "It was the beginning of her being able to start grieving over the loss of this daughter that she had never really done very well, throughout the remainder of her life."

Hearing Klotz talk about that woman makes me think about my mother. When I was 14, my maternal grandmother was killed in a car accident a day after she and my mother had argued. My mother felt guilty for years afterwards. I have no idea if she tried to resolve her guilt. Early on, Alzheimer's robbed her of speech. Late in the disease, it took away her ability to move. In the nursing home, she did neither, which means that her thoughts and feelings were very much trapped inside her.

When it works, Validation has other tangible benefits. Agitated people with dementia begin to express things they have buried for decades. They calm down. They become far less agitated.

They interact more with care providers and with family. They are less likely to withdraw into further stages of disorientation.

In most nursing homes that I have visited, some residents are quite prone to agitation and even violent behaviour. I ask Mandy Knight, the associate executive director at Country Meadows, if they ever have to use restraints on an agitated resident.

"No," says Knight.

"Ever?"

"No."

"How is that even possible?" I ask.

"I've never even imagined using them," she says. "I think because we're trained to know when to back off of residents and to leave them alone."

Proponents of Validation cite mountains of similar anecdotal evidence. It's a lot harder to find a study that *validates* the practice. A fairly thorough review of the evidence published in 2003 concluded that "there is insufficient evidence from randomized trials to allow any conclusion about the efficacy of Validation therapy for people with dementia or cognitive impairment."

Another critic faulted Feil for inaccurate descriptions of other therapies, and for confusion about these other approaches, but added tellingly that "these inaccuracies significantly detract from an otherwise engaging examination of Validation therapy . . . [which] is especially effective at humanizing older adults living with dementia."

Criticize it if you will, but decades after Feil first planted the seeds, the Validation approach is used by over 10,000 agencies in more than 16 countries. Validation may not teach care providers to be kind and empathic to people with dementia, but it gives them a method that makes it easier to understand what makes people with dementia agitated as they approach the end of their lives.

And it attracts extremely empathic people like Lorrie Quick, the young care provider I observed chatting about naps with a resident named Dorothy. The method seems to give them a way to tap into their natural abilities, overcome their fears, and accomplish almost unfathomable acts of bravery and kindness.

Lorrie Quick got a job as a personal care associate for people with Alzheimer's disease seven years ago. She was 21 years old at the time.

Like all new hires, Quick learned the Validation method as part of her training. But she didn't have to be taught how to tune into others. She says she was born that way.

"I can't remember *not* being that way," she says. "I had a best friend who lived down the street. When she would be sick, I would know. I pick up on another person's energy, feelings, vibrations, body language, all the ways that people speak without speaking. I can see the things that aren't there."

That includes being clairvoyant, an ability she paid little attention to until a terrible incident at her high school.

On Thursday, April 24, 2003, James Sheets, a 14-year-old student, boarded a school bus bound for Red Lion Junior High School, located on the east side of Pennsylvania's York County. During the bus ride, he listened to Limp Bizkit, the rap rock band from Jacksonville, Florida. Just after 7:30 A.M., the bus pulled up to the school. Sheets made his way to the glass-walled cafeteria and sat down at one of the tables.

Without saying a word, he stood up, pulled out a .44-calibre Magnum, and fired two times, striking Principal Eugene Segro once in the chest and killing him. Moments later, the young teen pulled out a .22-calibre revolver and shot himself in the right temple. He died instantly. No other shots were fired, and no one else was injured. The murder–suicide sent shocked students fleeing for safety, screaming as they ran from the building.

Standing outside the glass-walled cafeteria transfixed by it all was 14-year-old Lorrie Quick. "When everybody else ran, I watched," she says. "I just took it all in. That was the time that it all clicked in my head, that it made sense. I realized that I was different because I feel things differently than others."

Quick was different because, to her, the shooting wasn't happening for the first time. She had already seen it happen.

"I see parts of things, puzzles, pieces," she says, groping for the right words. "It's more like déjà vu, like I've already done something. Not when I'm awake, but sometimes, when I sleep. It's more about things that could happen or might happen."

Earlier that morning, before the shootings, she had woken up with a feeling of dread. "I had a feeling that it was going to be a bad day so I dragged my feet as long as I could," she recalls. "I begged my mom not to make me go to school, but she wouldn't let me. I purposely missed the bus, but my mom drove me into school anyway.

"She dropped me off at the door. I walked in the doors, then in the other door, and I made a right turn. You've got to understand, our high school cafeteria is nothing but glass windows from the exterior. Then, everything started happening."

A sense of déjà vu is not the only reason why Quick stood there watching instead of running away. By age 14, she knew she had an ability to tune in to others.

Before the day he shot Mr. Segro, Jimmy Sheets had talked about committing a violent act. "He never said a day or a time or a place that this was going to happen," she says. "We had no warning, no idea. We only had his comments that everybody laughed off."

But then, Quick had her dream. And as she looked at the carnage in the cafeteria, she realized she knew more about Jimmy Sheets than she had allowed herself to believe.

"I could see that he meant the things that he had said," she says. "Everybody thought that it was a joke. That's when I knew that I was different. If maybe I had realized that I understood before the situation happened, and that I was different. Maybe if I had my defining moment before then . . ." Her voice trails off before finishing the sentence.

"I was just a kid."

For two days following the shooting, Red Lion Junior High School was shut down, as authorities dealt with the aftermath of the traumatic events. Lorrie Quick says she shut down emotionally a lot longer than that.

"I didn't know how to handle it," she says. "I didn't know how to cope with it and I refused therapy. I didn't want any help. And anytime somebody would talk to me, it was as if I would just rip their head off. That's how I deal with things that I need time to process."

It took until well past her fifteenth birthday for Quick to be more like her usual open and empathic self. By then, she knew that she was different because of the way she had watched the dual shooting instead of running away. And, she also believed she was meant to do much more than watch. She was meant to act. But she didn't know how.

She found her answer at Country Meadows, thanks to Validation.

* * *

When the first workers at Country Meadows started taking training in Validation, Naomi Feil taught something on day one that stuck with Stephen Klotz, a technique Feil calls centring. At first, Klotz thought it was "a bit of New Age mumbo-jumbo." But the more Feil explained it, the more it made sense.

"If you're unfocused, Validation will not work well," Klotz recalls Naomi Feil telling the class. "If residents get any sense that your mind is somewhere else, distracted by other things, or too busy for them, they're going to sense that. And they are not going to open up and connect with you."

Klotz says that centring has three basic steps.

"Number one: slow down your pace, especially if you're a busy person in life," he says. "Number two: relax your body. That's important because there's a tendency most of us have when we're working, especially in stressful or busy jobs, to tighten up some-where on our bodies. Older people with dementia pick up on that. Number three: clear your mind."

Lorrie Quick didn't think much of the rest of Validation training because to her it seemed instinctively obvious. But centring was something new, something she figured she could use in her personal life. "I didn't know I could learn techniques to shut everything else out," says Quick. "[At the time] I didn't know how to handle my feelings, and centring taught me how to react to what I'm seeing."

Shortly after getting trained in centring, Quick had a chance to use it under life-and-death circumstances. When she was hired at Country Meadows, she was still working part-time at a convenience store nearby.

"The girl that usually works the night shift had been robbed and beaten, and they took the money," Quick recalls. "They had to take the girl to the hospital. She never came back to work."

The owner asked Quick to work the next night. She says she had a bad feeling that she was going to get robbed, but she went to work as requested at midnight. After her shift was under way, a tall man entered the store.

"I had a gun in my face and he's pointing it at me, and I had some other person in the back," recalls Quick. "But I wasn't really scared. I had customers in the store. I had responsibilities to take care of. I decided that freaking out and crying were not going to help the situation."

In that moment, she says she remembered standing outside the glass-walled cafeteria, watching Jimmy Sheets shoot and kill Eugene Segro before killing himself. Quick also remembered her Validation training. She began to centre herself.

"I literally breathed from my feet up," she says. "I planted my feet on the ground, took a deep breath, and I gave myself a little pep talk. I thought to myself that I can do this."

She remembered being taught how to mirror people with dementia.

"I mirrored the gunman," she says. "He said, 'Give me all your money' and I said, 'I can't open the cash.' I asked him why he was doing this. Then he demanded that I open up the safe. I told him that wasn't going to happen because I didn't know how to open it. Then I asked what led him to this moment. I must have struck a nerve or something because he yelled at the other guy and they left," she says.

Quick and the customers were unharmed, and no money was taken.

Later, when the police reviewed the surveillance video, they remarked that Quick remained calm and collected throughout the encounter and that her hands never shook. Quick says this was because she expressed curiosity about and interest in the gun-wielding robber.

"There is a difference between just listening and being empathic," she says. "You listen to respond to someone. If you're being empathic you're actually paying attention. You're invested and focused. You're in tune with that person. You feel what they feel. That's the difference."

Do not read Quick's story and assume that you suddenly have the gift of being able to talk down a gun-wielding assailant with soothing words. Validation wasn't designed for all-night store clerks. That said, the technique has now been adapted for paramedics, who sometimes encounter violent patients. And on that night, it may have saved Quick's life.

"Maybe I had to watch Jimmy Sheets to see how to talk my way out of that one," says Quick. "Had Jimmy not done what he'd done, maybe I wouldn't have made it out of that store that day. Everything happens for a reason."

That night, everything happened for a reason for Lorrie Quick, just as it did decades earlier for Naomi Feil—two women whose loneliness and trauma in childhood gave them unparalleled abilities to peek inside the troubled souls of others.

CHAPTER ELEVEN

Epiphany

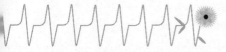

After several months of waiting, Philip Jackson contacts me to say he's got the results from my fMRI brain scan. He and Mathieu Gregoire—the lead investigator on the study in which I participated—scanned my brain while showing me photographs of people in pain. As per the study's protocol, I had to estimate how much pain the people in the photos were feeling just by looking at their facial expressions. The purpose of the fMRI scan was to see how much the emotional empathy parts of my brain light up when I see photos of people in pain.

As I've mentioned, my first reaction about getting any kind of test result is to feel nervous. Performance anxiety has been a part of my life from kindergarten. More than that, I've been primed for nasty surprises ever since Western University psychology professor Derek Mitchell told me I score high on the scale of Machiavellian tendencies—a notable empathy killer that is part of psychology's dark triad.

"What does my scan show?" I ask Jackson.

"It shows that you have a brain," says the neuroscientist.

At another time, I'd appreciate Jackson's sense of humour. I might even credit him with empathizing with my anxiety by trying to defuse it. But right now I'm too nervous to appreciate the gesture.

"I have a functioning brain and I obviously don't have a brain tumour," I reply. Brain tumour neurosis is not unique to me; at one time or another, almost every medical student I've ever known believes they have one.

"When we do scans like this, in the consent form, we explain to people that this is a research scan only," he says, conscious of the fact that I'm hanging on his every word. "We're not going to have this scan read by a radiologist. If we were to find something, there's a procedure to follow up. But, as researchers, we're not looking for any of that."

"Understood," I reply.

Jackson shows me a page containing three images of my brain taken during the same experiment but from three different angles: the top and side of my head and my forehead.

Peering at the images, I don't need a radiologist to tell me that I don't have a brain tumour. I can see that for myself.

These three images are what researchers call baseline measurements. In this experiment, they're designed to show that I followed instructions, and looked carefully at the photos of people whose pain I had to estimate, and did not stare off into space. Doing the latter would have rendered my test results useless.

Jackson tells me the scans show that I did what I was supposed to do.

Before he shows me the next set of fMRI images, Jackson wants to make sure that I understand a couple of important details about the study results. He reminds me that I looked at dozens of photos of people in pain while the fMRI machine scanned my brain; for control purposes, I also looked at faces of people who weren't in pain.

The first detail Jackson wants me to understand is that the scans he's about to show me reflect an *average* measurement of my brain activity while viewing every photo, not any one of them. The second detail is that the scans do not show what parts of my brain light up when I'm empathizing emotionally with people in pain; the images show the *contrast* in brain activity (in the empathy parts of the brain) between seeing photos of people with pained looks on their faces and seeing photos of people who aren't in pain.

With those caveats in mind, he hands over the second page of three scans. To me, they look different from the initial control set. Several areas of my brain appear brighter and larger than they did in the first set of images: the anterior insula cortex, the dorsal anterior cingulate cortex, and the anterior midcingulate cortex. These just happen to be the very same areas of the brain that Jackson's mentor Jean Decety found to light up "when one experiences pain and when one perceives, anticipates, or even imagines pain happening to others."

This seems promising. I press Jackson for more details. "From the images you're showing me, would you say that when I looked at photos of people in pain, my right anterior insula and my dorsal anterior cingulate cortex were activated as expected, less or more?"

"All I can say is that they were activated during your observation and assessment of people in pain, in contrast with when you observed people with neutral facial expressions," Jackson replies.

I ask if he can tell me how my scans compare to those of other people.

"fMRI is not used reliably to provide individual data," says Jackson. "I cannot say anything about how the relative activity in your brain compares to that of another group."

"How do I compare to other health professionals?"

"This is very tentative but one might say that you were less affected by repetitive pain exposure than people who are not health care professionals," he says.

Even though the emotional empathy parts of my brain lit up, the effect was not as great as it would be for people who aren't doctors or nurses.

Jackson explained his hypothesis the first time we met on the day I had my scan. He and other researchers are uncovering evidence that the longer health professionals like me practise, the more numb we become to the pain our patients feel. If Jackson's hypothesis is correct, my 35-year career in the ER may be making me less empathic than I was when I started out.

The conclusions are preliminary, and Jackson won't commit to them just yet. "If this pattern is replicated in a group of health care professionals, it might suggest that they show some form of habituation to people in pain, which I would argue is adaptive to the job," he says.

By "adaptive," he means being able to function in the ER without being burdened emotionally by the suffering of patients in pain.

Jackson advises me not to jump to conclusions about how much I care about my patients. Underestimating the level of pain that my patients feel does not mean necessarily that I ignore their pain or manage it inadequately. Still, he is intrigued enough that he has applied for funding to the Canadian Institutes of Health Research, one of Canada's most respected granting agencies, to lead a study based at Laval University and involving a larger group of health professionals.

The results leave me disturbed that being a health professional may have damaged my capacity to be kind. "Can you say
that I'm an empathetic guy?" I ask.

"From your brain scan, I can say that you seem to have the
expected response to the pain of others, which we think is one of
the many processes involved in empathy," he replies.

He also says this:

"From the persistence with which you ask the same questions, even after I have provided several answers, I would say
that you might have some difficulty putting yourself in the shoes
of others—me in this case."

That stings a bit. I get why he's uncomfortable. In scanning my
brain and including me in one of his studies, Jackson is sticking his
professional neck out. He is rightly concerned that I'll misrepre-
sent the scan as, in his words, a "revolutionary empathy detector."

Still, as Derek Mitchell pointed out, I also have Machiavellian tendencies.

"Why can't the scan show if I'm empathic?" Even after
Jackson's gentle teasing, and his insinuated request that I stop,
I'm still asking questions.

"Empathy is a combination of affective [emotion], cognitive
[thinking], and behavioural [doing something] responses," he
says. "One fMRI study usually focuses on one component at a
time. Future studies that can test multiple aspects of empathy at
the same time will provide a more complete picture."

I know I've been pressing hard for answers. I have one more
question to ask.

"What is the best way for a curious person like me to find out
how empathic I am? Is there some other brain scan I can do?"

For a neuroscientist who works with advanced brain imaging
technology, his answer surprises me.

"I would say the best way to know if someone is empathic or not is to ask people that interact with them on a regular basis," he says. "There is data showing that we tend to be more empathic toward people close to us. So, if these people find that we are not empathic, chances are that this would generalize to other people as well."

Asking my family what they think of my empathic capabilities is an excellent idea that had not occurred to me.

Philip Jackson isn't just smart; he's wise.

Tamara Broder is a career advisor and employment counsellor with the YWCA of Toronto. Since moving to Toronto from Winnipeg in 1995, she has helped hundreds if not thousands of women, and more recently, men, brush up on their office skills, fix their resumés, and find new jobs.

Tamara is one of the most empathic people I know. She also happens to be my partner since 1996. Almost every day, she comes home with heartfelt stories that often make me cry.

"I met a client just the other day," she says, running her fingers through curly dark hair. "She lost her baby when she was eight months pregnant, and three months later, her husband died. I meet people all the time who have gone through terrible hardships."

Empathy is a big part of what she tries to bring to her working relationship with clients.

"Sometimes, I need to remind myself to be more empathetic, to put myself in someone's place and not to judge them," she says. "Life can change so quickly, and when it does, it has such a snowball effect, with one circumstance leading to another. Everybody's got a story, and I feel honoured and privileged to hear people's stories, and that they feel comfortable sharing them with me."

Like Shalla Monteiro, the woman who befriended Raimundo Arruda Sobrinho, a homeless poet on the streets of São Paul, Tamara has an uncanny ability to inspire people to confide in her. Like Mark Wafer, the entrepreneur whose success in the donut business came largely from hiring workers who are disabled, Tamara helps employees trying to make a fresh start. But unlike the donut maven, she has rarely if ever been where her clients are. Her empathy doesn't come from having been disadvantaged herself.

Still, when I ask her where she thinks her empathy comes from, she says it comes from knowing what it's like to be told you're not good enough. "When searching for a job, people don't feel good about themselves," she says. "It's very frustrating and lonely. For people who were let go or laid off from a company where they've worked for years, they feel a sense of anger, betrayal, or hurt. So, I always try to say something positive about them and about their skills and abilities."

She tells me about a client she saw recently. She says she told the woman that it might take some time to find the right job, but that she is "very employable."

"She was so grateful for that," Tamara says. "She said she had stopped believing that to be true. It's about saying something positive, validating their feelings, or just telling them that they're doing a good job. I also tell them I'm happy to help them create a plan to find a job, work on their resumés, and any other steps they need to take. I think people appreciate that I'm here to help."

I've kept Tamara up to speed about my search around the world to learn about empathy from the kindest people I can find.

She's heard about everyone I've met, from Mark Wafer to Shalla Monteiro. Tamara was with me in Tokyo when roboticists Tadashi Shimaya and Takeshi Mita introduced us to Asuna, the android whose expressive eyes and facial movements are pitch perfect for a 15-year-old girl.

I told her about my fMRI brain scan, my somewhat frustrating attempts to pry some definitive answers from Philip Jackson, and his suggestion that I ask a family member instead.

"I think you're one of the most empathic people I know," says Tamara.

"What makes you say that?" I ask her.

"If empathy is the ability to put yourself in someone else's shoes, and feel what they feel, or imagine what they feel, you do that in spades," she says. "Just the other day, you did that with our children. You can put yourself in their place right away. It stands out about you. You have a skill. You have a talent. You have a unique gift to do that."

When we first started dating, I had been working as an ER physician for close to 15 years. I want to know if Tamara thinks working in health care for close to 35 years has caused me to lose my capacity to be kind.

"I hear you say that there are patients who drive you crazy," says Tamara. "They bug you, and you get pissed off at them. Well, I get pissed off at clients too. But I see you. You genuinely care about the patients you see, and that's a wonderful quality to have after 35 years."

As nice as it sounds, hearing Tamara talk about me like that makes me uncomfortable. I want to know if she's seen me be unkind.

"You're not always empathic when you're driving, where your competitive nature comes out," she says. "You've talked about the accolades that a colleague has received lately. That's being competitive, but it can also be a case of envy. I don't think there's anything wrong with asking why someone else gets more awards than you. Very few people are naturally empathic. For most of us, it's a constant battle between our good side and our not so good side. I think most people have to work at being empathic."

Perhaps feeling deeply disappointed at not being singled out for awards or special praise has given me empathy for others who have suffered disappointment.

I tell her that Derek Mitchell discovered that I have Machiavellian tendencies.

"Don't most people have Machiavellian tendencies?" Tamara asks.

"That's the thing," I reply. "It's not that I have a Machiavellian personality disorder, but that I have a cluster of tendencies. I'm above average. Does that bother you?"

Tamara explains that her empathy has limits too. She starts talking about the things she and her colleagues have to do to maintain funding from the province for their employment program. Every month, they face pressure to show good outcomes, that clients are getting jobs, for instance, and that files are being closed.

"Sometimes, we just have to get the outcome," she says. "We keep calling the client, and this person isn't calling us back. Sometimes, we talk about clients not in an unkind way—but in an exasperated way. We have to get the outcome, and even if we're bugging somebody, we still have to do it. So sometimes the ends justify the means."

"Have you ever seen me act mean?" I ask her.

"You rant about people you don't like," she says. "Sometimes, you're mean to yourself. You are your worst critic. You tear yourself apart sometimes."

"I don't know otherwise," I reply.

"I know you don't," she says. "But that can always change."

And with that, Tamara kisses me on the cheek, gets up, and walks away.

My partner says I'm empathic, even after 35 years in the ER. According to Philip Jackson, my brain scan shows that a long career in the ER has dulled my capacity to empathize emotionally with patients in pain, but it hasn't obliterated it. Derek Mitchell says I have a Machiavellian streak, but he also says my empathy is in the "cuddly" range.

There's something familiar about Jackson's suggestion to ask Tamara if I still have empathy. When I was a kid, *The Wizard of Oz* was one of my favorite films. It's one of the happiest childhood memories I possess. Nearly half a century before there were PVRs, *The Wizard of Oz* was broadcast just once a year. On that special occasion, my mother would sit down with my sister and me to watch it.

There was the Wizard handing a diploma to Scarecrow, a ticking clock in the shape of a heart to Tin Man, and a medal to Cowardly Lion. Each is a symbol of something they already have. They just didn't realize it.

I never lost my empathy to begin with. I've searched for it around the world. It was inside me all the time.

In writing this book, I have travelled near and far on a journey to learn about kindness. I have met many people whose empathy inspires me.

Some of us are born extraordinarily kind. But most get there only after experiencing pain and then learning from it. Naomi Feil was a lonely child living in a nursing home, where she was befriended by an elderly woman named Florence Lew. What Feil taught me is to accept friendship, with gratitude and without judgment, wherever it comes from. Feil's loneliness was the engine with which she connected with Lew, gaining a deep understanding of people with dementia in the process.

Mark Wafer overcame many obstacles as a child born deaf and grew up into a man who helps people with profound disabilities find meaningful employment. He showed me how to use the pain caused by taunts and humiliation to identify with others in similar circumstances.

Theodore Fontaine, the Indigenous elder and writer, endured unimaginable abuse at an Indian residential school. For him, the most painful aggression he experienced was the obliteration of his cultural heritage. He taught me that sometimes the surprising way to regain one's identity is not to forget the past, but to remember it warts and all.

Fontaine also taught me that empathic people accept emotional connection from whoever extends it in whatever form it takes. He doesn't want exclusive jurisdiction over pain. Instead, he seeks the commonalities between his suffering and that of others, no matter how trivial by comparison.

Shalla Monteiro was born gifted in empathy. But it took the bitterness of her parents' troubled marriage to learn to live in the moment and to find moments of bliss and of deep connection with others.

Like Shalla, Paul Mackin, the bartender at Ground Zero, is gifted in empathy. But it took the crisis of 9/11 for him to change his career and his life by turning the bar closest to Ground Zero into a place of kindness for first responders.

Lorrie Quick taught me that sometimes, empathy comes with a predetermined destiny, to accept it and to prepare for it.

Mary Gordon, the creator of Roots of Empathy, has boundless compassion for others. From her, I learned the lesson her parents taught her decades ago: rich or poor, advantaged or disadvantaged, we are all the same. Though her course has been given to 1 million children and counting, she told me that no course alone can train us to be empathic; what it can do is teach us to notice what it looks, sounds, and feels like to be kind.

Derek Mitchell tested my personality and told me that while I have empathy in spades, I also have a tendency to sometimes prioritize my own needs to the detriment of others. Mitchell also reassured me that humans are complicated creatures, made up of competing wants and desires. Now that I'm aware of them, I can keep watch over them.

Philip Jackson scanned my brain for empathy but told me to look instead inside my own heart.

Each of them and others helped me realize that if you want to be kinder, first look for answers inside you.

If you want to be kind to others, the first person you must be kind to is you.

If you want to connect with others, the first person you must connect with is you.

I believe that most of us have one kind of pain or another tucked away in a corner of our hearts. Denying or running away from it won't make you feel better, but it will make it harder to connect with others. If you can, take a risk and embrace your flaws or what you might see as your less than perfect side.

Christian Keysers, the researcher who helped discover mirror neurons in the brain, taught me that humans are hard-wired to be empathic and kind. But he also taught me that empathy is not a biological obligation but a choice made by weighing pros and cons.

The choice is yours.

Acknowledgements

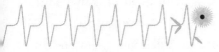

The book you are reading is very different from the one I first proposed to my publisher. I originally set out to examine the challenge facing medical staff of being empathic toward patients in today's health care environment. Jim Gifford, my editor at HarperCollins, made a counterproposal: that I should aim higher by examining empathy and kindness in the world at large.

While I appreciated the vote of confidence, I was scared witless at venturing outside the walls of the hospital and far outside my comfort zone. For one thing, aside from common sense, I knew little or nothing about the neuroscience of compassion. If I explored kindness (or lack thereof) in health care, I'd be on familiar turf. To write a book about kindness at a donut shop or a bar in Manhattan meant surrendering the advantage in credibility I've worked to build over many years as a physician and as a medical writer and broadcaster.

To write the book that Jim Gifford was looking for, I'd have to rely on my natural curiosity and the help of many hands. My goal was to meet some of the kindest people on the planet, to spend some time with them, and to learn and pass on to you what makes them exemplars of empathy.

Fortunately, I had not one but two extraordinarily smart angels without whom my search would have been long and fruitless. Karin Chykaliuk is a talented producer I first met at the CBC. For many years, Karin has been the lead producer on many live events put on by CBC Radio. These include town hall meetings in which episodes of *White Coat, Black Art*, *The Current*, and other CBC Radio shows are staged in front of a live audience. Karin has also been the producer of CBC Toronto's *Sounds of the Season*, an annual event that includes live radio and TV broadcasts, performances by acclaimed local musicians, and meet-and-greets with CBC local and network stars. The event raises phenomenal amounts of money for local charities.

Karin had never worked on a book before but jumped at the chance to dig out some incredible stories. Without Karin, I would not have met Paul Mackin or gone to O'Hara's Restaurant & Pub, the only bar inside 9/11's Ground Zero. Without Karin, I would not have met Mark Wafer, the donut store entrepreneur who employs many people with physical and emotional disabilities.

My second angel is Erin James Abra, a talented writer and editor I met when she was a postgraduate student in the science journalism program at Ryerson University. Erin did essential research on my last book, *The Secret Language of Doctors*. For *Kindness*, Erin assembled masterfully a reference list of the latest scientific research on empathy. She also reached out to top researchers in Canada and around the world. Without Erin, I would not have been able to persuade Philip Jackson, a top neuroscientist at Laval University in Quebec City, to do a functional MRI scan of my own brain to see if I still have the right circuits for empathy.

Erin did a lot more than that. As I finished chapters of the book, she became my muse and my most astute critic. She told me straight up when my scientific explanations didn't add up and when my stories didn't work. Her pointed but polite suggestions always made the book better. And then, when she was in her eighth month of pregnancy, Erin rose to the challenge yet again, helping me cut four chapters and 40,000 words from a book that was at risk of losing its focus. That makes Erin the proud mum of a healthy baby boy *and* the skilled midwife of a book about kindness.

As part of my travels in the preparation of this book, I visited Brazil and Japan. In both countries, I depended heavily on the help of fixers. These are local journalists who help arrange interviews and other logistics and act as translators in the field. In Japan, I had the pleasure of working with Chie Matsumoto. Chie worked as a reporter for the *Asahi Evening News* and the *International Herald Tribune/Asahi Shimbun* and as a Tokyo correspondent for the German Press Agency, dpa. She became an independent journalist and fixer in 2009. Chie writes on labour issues and problems with nuclear power plants.

In Brazil, my fixer was Kate Steiker-Ginzberg, a freelance producer who worked extensively with CBS News during the 2016 Summer Olympics in Rio de Janeiro and during the Zika outbreak.

Both Kate and Chie also acted as fixers when I reported for CBC Radio's *The Current* on the anniversary of the Fukushima Daiichi nuclear disaster and on the Zika outbreak in Brazil. Their help enabled me to spend more time absorbing the culture in both countries. Special thanks go to *The Current*'s executive producer, Kathleen Goldhar, and producers Joan Webber and Lara O'Brien, who were instrumental in helping bring those stories to air.

A big thank you goes out to my colleagues at *White Coat, Black Art*. Jeff Goodes produced my interview with Validation creator Naomi Feil. His sensitive production of her interview and my visit to Country Meadows to see Validation up close won our show a 2017 Gabriel Award for radio. I also want to thank senior producer Dawna Dingwall for her support as I researched and wrote the book.

I benefited from the kindnesses of many people as I researched this book. I want to express my gratitude to those who told me their stories and who answered patiently all of my questions.

By far the most noteworthy acts of kindness were extended by the men and women who helped me obtain an on-the-spot visa to enter Brazil when all hope appeared to be lost. Special thanks go to Marcela, the consular official who returned my call at 2 A.M., and who next day opened up the office of the Consulate General of Brazil at noon on the Saturday of the Easter weekend. I also want to thank Ricardo, the official on call at the Embassy of Brazil, who also answered my call in the dead of night and gave me Marcela's phone number; Evan, the photographer who opened up his studio to take my passport photo at 4 A.M. and Lynda Shorten, my manager at CBC, who drafted and signed a letter that I was entering Brazil on CBC business at 8 A.M. If any one of them had not stepped up, I would have had to cancel my entire research trip.

Thanks go to Natalie Meditsky, who as production editor guided the book through its final stages. To Jim Gifford, my editor and editorial director for non-fiction at HarperCollins, heartfelt thanks for being so patient and supportive as I researched and wrote this book and for not complaining when I asked for extra time. Anne Holloway did a great job copyediting the book.

To my agent, Rick Broadhead, a big thank you for representing me and especially for acting as my verbal punching bag when times got tough. You earned your pay, and you were kind about it too.

My parents, Sam and Shirley Goldman, were alive when I wrote my first two books. *Kindness* is the first book I've written since they passed away. Both of them taught me about empathy, each in their own unique ways. For 15 years, Dad was our mother's primary caregiver after she was diagnosed with Alzheimer's disease. He taught me that love is not what you say but what you do; his kindness and tenderness with Mum awe and inspire me even more than they did when my parents were alive. His steadfastness lives on in my sister, Joanne Orliffe. With a little help from Naomi Feil, my mother taught me to see the person who still exists in spite of dementia.

My most heartfelt thanks go to my children, Kaille and Alex, who remind me constantly how important it is to see the world from their point of view. My beloved partner, Tamara, is the most empathic woman I know. Without her love, encouragement, and support, this book would not be possible.

Brian Goldman, MD
November 12, 2017

Index